Innovative SAP SuccessFactors Recruiting

A Guide to Creating Custom Integration and Automation

Anand 'Andy' Athanur
Mark Ingram
Michael A. Wellens

Apress®

Innovative SAP SuccessFactors Recruiting: A Guide to Creating Custom Integration and Automation

Anand 'Andy' Athanur
San Francisco, CA, USA

Mark Ingram
Edmonds, WA, USA

Michael A. Wellens
Davidson, NC, USA

ISBN-13 (pbk): 978-1-4842-7424-8
https://doi.org/10.1007/978-1-4842-7425-5

ISBN-13 (electronic): 978-1-4842-7425-5

Managing Director, Apress Media LLC: Welmoed Spahr
Acquisitions Editor: Divya Modi
Development Editor: Mark Powers
Coordinating Editor: Divya Modi

Cover designed by eStudioCalamar

Cover image designed by Freepik (www.freepik.com)

Distributed to the book trade worldwide by Springer Science+Business Media New York, 1 New York Plaza, New York, NY 10004. Phone 1-800-SPRINGER, fax (201) 348-4505, e-mail orders-ny@springer-sbm.com, or visit www.springeronline.com. Apress Media, LLC is a California LLC and the sole member (owner) is Springer Science + Business Media Finance Inc (SSBM Finance Inc). SSBM Finance Inc is a **Delaware** corporation.

For information on translations, please e-mail booktranslations@springernature.com; for reprint, paperback, or audio rights, please e-mail bookpermissions@springernature.com.

Apress titles may be purchased in bulk for academic, corporate, or promotional use. eBook versions and licenses are also available for most titles. For more information, reference our Print and eBook Bulk Sales web page at http://www.apress.com/bulk-sales.

Any source code or other supplementary material referenced by the author in this book is available to readers on GitHub via the book's product page, located at www.apress.com/978-1-4842-7424-8. For more detailed information, please visit http://www.apress.com/source-code.

Printed on acid-free paper

To Kim Lessley.

Table of Contents

About the Authors

Anand 'Andy' Athanur has over 25 years of HR systems experience and currently works at SAP enabling partners to innovate their solutions with SAP SuccessFactors.

Mark Ingram has extensive knowledge of recruiting systems and integrations across his 25-year career as a developer, product manager, and consultant. There isn't a recruiting problem that he doesn't like!

Michael A. Wellens, M.S., is a certified SAP SuccessFactors consultant with over 15 years of human resources information systems implementation experience. He has successfully launched a variety of core HR and talent management solutions across a number of Fortune 500 companies around the world and has written on diverse SAP, HR, and IT topics. You can follow him on LinkedIn or on Twitter.

About the Technical Reviewer

 **Rinky Karthik** is a Principal HXM Cloud Architect, part of the Global Solutions Architect team at SAP. She is a Certified SuccessFactors professional with multiple global implementations in various SuccessFactors modules and has vast experience in SAP On-Premise implementations. In her role as an SAP HXM Solution Architect, she has been instrumental in providing SuccessFactors Solution Architecture advisory services for complex, global organizations with strategies and innovative business models for a successful cloud transformation while considering leading practice guidance.

Rinky has co-authored the SAP Press book *Integrating SAP SuccessFactors* and authored the SAPinsider book *SAP SuccessFactors Recruiting: The Complete Integration Guide*.

Acknowledgments

Thank you to David Ludlow for starting Mark down the recruiting path.

Thank you to the SuccessFactors Recruiting product teams for their continued support over the years.

Introduction

Talent acquisition is a critical business process for any organization. Without key resources, organizations simply cannot function. The magnitude of this has been felt even more recently as the world continues to face the economic challenges associated with the global pandemic. As companies grapple with one another for skilled resources, they often look to their recruiting systems to help find a competitive edge. As many customers subscribe to cloud software such as SAP SuccessFactors, key differentiators become even harder to find. Cloud customers may ask, "How can we stand out among the community if many of us are using the same software?" Furthermore, in our experience, many vendors and customers approach SAP SuccessFactors and need resources to bridge the gap between product documentation and real-world examples.

In this book, we look at ways to innovate your SAP SuccessFactors Recruiting solution. We step through the end-to-end recruiting business process and highlight key areas where SAP SuccessFactors allows for automations and integrations. At each step, we walk you through some of the tools that can help enhance the process in your own innovative way. In Chapter 1, we conduct a basic overview of ODATA to get you familiar with the protocol that can be used throughout the recruiting process. In Chapter 2, we take a look at how to automate requisition creation using non-SAP systems. Chapter 3 looks at this same process using middleware. Chapter 4 examines how to enhance the recruiting posting process. Then in Chapter 5, we look at different ways to enhance candidate engagement. Chapter 6 looks at Robotic Process Automation as a way to automate a variety of recruiting processes. Likewise, Chapter 7 shows how such processes can also be automated using the Integration Center. As we progress to Chapter 8, we see how the system can integrate with assessment vendors using the Assessment Integration Framework. In Chapter 9, we focus on ODATA Integrations. Then in Chapter 10, we see how SAP SuccessFactors can integrate with background-check vendors using the Background-Check Integration Framework. Chapter 11 walks us through how to help automate the candidate offer process using SAP SuccessFactors Business Rule functionality. Then in Chapter 12, we look at how the Intelligent Services functionality can help automate processes. In

Chapter 13, we look at how we can automate the hire process with non-SAP systems. Lastly, in Chapter 14, we wrap up with how to realize business value and draw conclusions. After completing the book, you should have a solid understanding of what tools are available to enhance your recruiting system at each step. In addition, you will have an understanding of practical examples from the authors' real-world experiences.

CHAPTER 1

Introduction

You are about to learn the basics of retrieving SAP SuccessFactors data using free-to-use tools. In this chapter, we will lay the foundation of the rest of the book by connecting to SAP SuccessFactors and retrieving some job requisitions. We'll be using ODATA and the REST protocol. This chapter will begin by describing SOAP web services vs. ODATA and REST.

What Is ODATA?

The SAP SuccessFactors Human Experience Management (HXM) Suite supports extensibility and integration with third-party solutions using comprehensive application programming interfaces (APIs). These APIs are used to access data across the suite. The APIs are based on the ODATA protocol and offer methods for CRUD (Create, Read, Update, and Delete). The ODATA protocol, originally developed by Microsoft, offers a standardized way to access APIs. The ODATA protocol replaces the SFAPI SOAP web services.

How Do SOAP Web Services and ODATA Differ?

SOAP is a messaging protocol that's used to exchange structured information via web services. Web services perform a specific purpose and are designed around a process. In recruiting, RequestBackgroundCheck may be a service. It is asynchronous and has an expected structured response such as Background Order ID. The content of the message exchange is called the payload. Payloads have been driven by HR tech industry standards such as the HR-XML protocol. SOAP APIs have an abstraction from the particular software database, so software applications can communicate without being aware of each other's data model.

© Anand 'Andy' Athanur, Mark Ingram and Michael A. Wellens 2022
A. A. Athanur et al., *Innovative SAP SuccessFactors Recruiting*, https://doi.org/10.1007/978-1-4842-7425-5_1

ODATA REST services are lightweight and used for direct interaction with database tables. They are stateless, meaning communication sessions with asynchronous back and forth are not supported. Every request results in a simple response.

Data Model Navigation

ODATA entities can be thought of as database tables. For online stores, a Customer entity is tied to Orders, which in turn are tied to Products. In SAP SuccessFactors Recruiting, Job Requisitions are tied to Applications, which in turn are tied to Candidates.

Whereas SOAP web services are designed to be read by techie humans, using XML elements such as RequestInterviewResults that are recognized across the HR tech industry, ODATA directly represents the data model, using a generic way of representing entities.

Using a web service RequestInterviewResults might pass a job requisition ID and return interview results for each candidate. The response might look something like this:

```
<InterviewResults>
  <InterviewAssessment>
    <Candidate>
  <CandidateID>1234</CandidateID>
  <Name>Charlie Brown</Name>
</Candidate>
  <InterviewDate>12/21/21</InterviewDate>
  <InterviewScore>A-</InterviewScore>
</InterviewAssesment>
<InterviewAssessment>
<Candidate>
  <CandidateID>4321</CandidateID>
  <CandidateName>Lucy</CandidateName>
</Candidate>
....... Etc.
</InterviewAssessment>
</InterviewAssessments>
```

With ODATA, we navigate entities using our request parameters rather than it being handled by predefined logic that drives the web service. Using the same example earlier, we could have the data model navigation as shown in Figure 1-1.

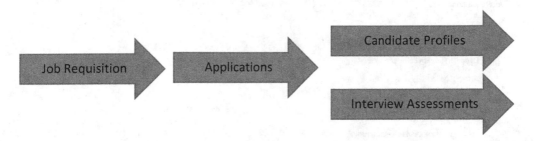

Figure 1-1. *Data Model Navigation from Job Requisition to Interview Assessment*

CRUD (Create, Read, Update, Delete) operations are performed on ODATA *entities*. There are hundreds of entities across SAP SuccessFactors. This can be both powerful and confusing. Fear not, documentation and this book are your friends.

In the SAP Help Portal, the following useful guides give further information:

- The SAP SuccessFactors HXM Suite OData API: Developer Guide (Dev Guide)

- The SAP SuccessFactors HXM Suite OData API: Reference Guide (Reference Guide)

ODATA Data Model Example

Now that we've described how ODATA works, let's look at the ODATA model by diving into the SAP SuccessFactors system. An interactive view of the data model can be seen within the ODATA Data Model. In the search bar at the top right of the screen, type "ODATA API Data Dictionary." Options will appear as soon as you start typing.

When the Data Dictionary screen appears, click *Entity* and then select *Tag* Recruiting (RCM) and *Name* JobRequisition. This will show the screen in Figure 1-2.

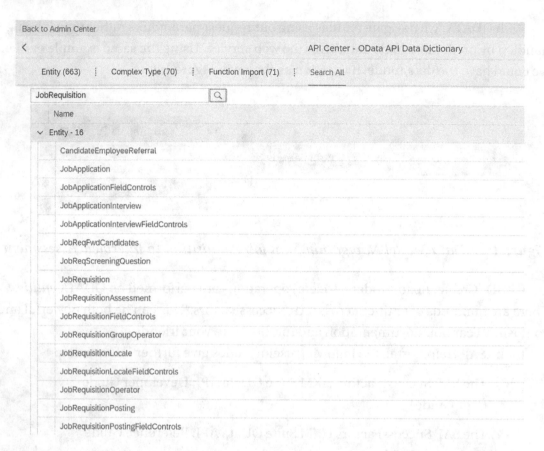

Figure 1-2. *ODATA Data Dictionary List of Entities Related to JobRequisition*

Select JobRequisition and the entity will be shown in detail, with the definitions of each field as shown in Figure 1-3.

Back to Admin Center

‹ API Center - OData API Data Dictionary

Entity (66) ⋮ Complex Type (70) ⋮ Function Import (71) ⋮ Search All

Tag: [Recruiting (RCM) ⌄] Name: [JobRequisition ⌄] Supported Operations: Query, Insert, Merge, Replace, Upsert, Delete

Property Name ▲	Label	Type	sap:picklist	Business Key	Nullable	sap:required
age	Age	long		false	false	false
appTemplateId	Application Template Id	long		false	false	true
assessRatingScaleName	Assess Rating Scale Na...	string		false	false	false
candidateProgress	Candidate Progress Sta...	long		false	false	false
closedDateTime	Date Closed	datetimeoffset		false	false	false
comment	Comments,Requisition ...	string		false	true	false
commission	Commission	decimal		false	true	false
corporatePosting	Corporate Posting	bool		false	true	false
costCenterCode	CostCenter Code	string		false	true	false
costOfHire	Cost of Hire	decimal		false	true	false
country	Country	string		false	true	true
createdDateTime	Date Created	datetimeoffset		false	false	false
currency	Currency	string		false	true	false
customString3	Job Level	string		false	true	false
customString4	Job Status	string		false	true	false
customString8	Job Pay Grade	string		false	true	false

Figure 1-3. *The Fields of the JobRequisition Entity in the ODATA Data Dictionary*

In addition to showing the fields within the JobRequisition entity, the supported operations are shown. These are Query, Insert, Merge, Replace, Upsert, and Delete. Not all operations are supported for all entities. The possible operations for an entity can be found in the ODATA Data Dictionary as well as the Reference Guide.

The second way that we will be navigating the data model is through Integration Center. This will be described in Chapter 2.

How to Execute Simple ODATA Calls
Prerequisites

We can quickly get a list of all entities while confirming that we can access the ODATA API using a username and password shared with us by a system administrator. You will also need the SAP SuccessFactors Company ID.

The endpoint URL to access the ODATA APIs depends on the data center housing your SAP SuccessFactors instance. We will be using a demo system for our examples. The endpoint is `https://apisalesdemo8.successfactors.com/`.

The data center URLs to access your data center can be found in knowledge base article (KBA) here `https://launchpad.support.sap.com/#/notes/2605498`. You will need an SAP user, such as an S-user to access this site.

To query a list of entities, the URL is <Endpoint>/odata/v2. Our URL is `https://apisalesdemo8.successfactors.com/odata/v2`.

You will be prompted for a username and a password. The username is *user@companyid*. The user should be an integration user with limited access to the system. If you are responsible for creating roles and users, read the section "Setting Up Required Users and Permissions."

Setting Up Required Users and Permissions

Integration users should have just enough permissions to make ODATA requests and nothing more. They should not have access to any front-end functions that a human user would perform. They should also have a password that never expires as an expiring password would cause a failure of any API calls that SuccessFactors receives.

A separate role-based permission (RBP) role and group should be created for the integration user. It's assumed that if you're reading this book, you know the basics of role-based permissions. Figure 1-4 shows the permissions recommended for a recruiting integration user. Not all ODATA permission may be needed. Foundation or MDF object permissions such as Location may also need read access.

Role Permission Section	Permission	Comments
General User Permissions	User Login	
	Login Method (Password)	If SSO is enabled
	Permission to Create Forms (check any requisition templates)	If requisitions are to be created using integration
Recruiting Permissions	Check off all ODATA permissions	

Figure 1-4. *Role Permissions for Recruiting Integration User*

If you are performing API testing from your computer and are the only person using that integration user, then an expiring password is not such a big deal. In a production environment, it is disastrous for integrations. An exception needs to be made to the password expiration policy.

To set a password expiration exception, type "Password" in the action search in the top right of the SuccessFactors screen. "Password & Login Policy Exceptions" will be one of the options. The search tool within Admin Center can also be used as shown in Figure 1-5.

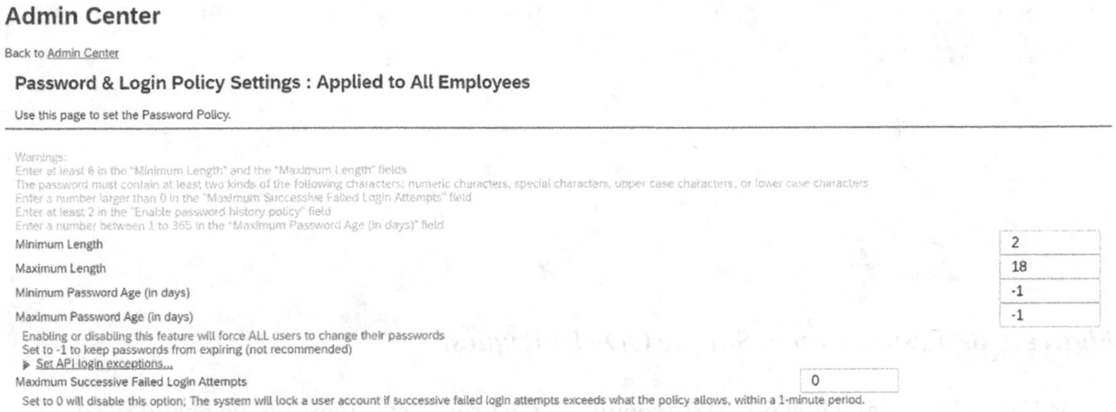

Admin Center

Back to Admin Center

Password & Login Policy Settings : Applied to All Employees

Use this page to set the Password Policy.

Warnings:
Enter at least 6 in the "Minimum Length" and the "Maximum Length" fields
The password must contain at least two kinds of the following characters: numeric characters, special characters, upper case characters, or lower case characters
Enter a number larger than 0 in the "Maximum Successive Failed Login Attempts" field
Enter at least 2 in the "Enable password history policy" field
Enter a number between 1 to 365 in the "Maximum Password Age (in days)" field

Minimum Length	2
Maximum Length	18
Minimum Password Age (in days)	-1
Maximum Password Age (in days)	-1

Enabling or disabling this feature will force ALL users to change their passwords
Set to -1 to keep passwords from expiring (not recommended)
▶ Set API login exceptions...

Maximum Successive Failed Login Attempts	0

Set to 0 will disable this option; The system will lock a user account if successive failed login attempts exceeds what the policy allows, within a 1-minute period.

Figure 1-5. *The Password & Login Policy Settings Screen*

Click "Set API login exceptions." You can now add an exception by clicking the "Add" button. Enter the integration username, and set the maximum password age to "-1" for no expiration or another period depending on the policy of your organization. An IP address restriction is required. This limits the IP addresses from which the user can make calls to your SuccessFactors instance. An example would be the IP address of an assessment solution calling SuccessFactors.

Calling SuccessFactors from a Browser

Open your favorite browser and type the ODATA endpoint for your SuccessFactors instance. For our demo instance, it's `https://apisalesdemo8.successfactors.com/odata/v2`. You will be prompted for a username and password. The username format is *username@companyid*. After clicking Sign In, you will see a list of all ODATA entities like the one in Figure 1-6. Congratulations! You have executed your first ODATA API request.

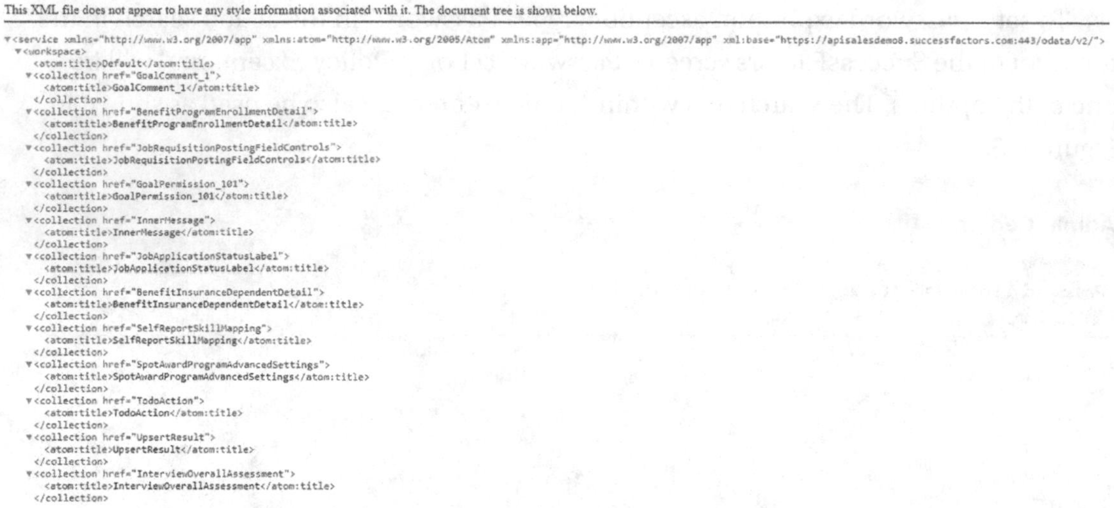

```
This XML file does not appear to have any style information associated with it. The document tree is shown below.
```

```
▼<service xmlns="http://www.w3.org/2007/app" xmlns:atom="http://www.w3.org/2005/Atom" xmlns:app="http://www.w3.org/2007/app" xml:base="https://apisalesdemo8.successfactors.com:443/odata/v2/">
  ▼<workspace>
     <atom:title>Default</atom:title>
     ▼<collection href="GoalComment_1">
        <atom:title>GoalComment_1</atom:title>
     </collection>
     ▼<collection href="BenefitProgramEnrollmentDetail">
        <atom:title>BenefitProgramEnrollmentDetail</atom:title>
     </collection>
     ▼<collection href="JobRequisitionPostingFieldControls">
        <atom:title>JobRequisitionPostingFieldControls</atom:title>
     </collection>
     ▼<collection href="GoalPermission_101">
        <atom:title>GoalPermission_101</atom:title>
     </collection>
     ▼<collection href="InnerMessage">
        <atom:title>InnerMessage</atom:title>
     </collection>
     ▼<collection href="JobApplicationStatusLabel">
        <atom:title>JobApplicationStatusLabel</atom:title>
     </collection>
     ▼<collection href="BenefitInsuranceDependentDetail">
        <atom:title>BenefitInsuranceDependentDetail</atom:title>
     </collection>
     ▼<collection href="SelfReportSkillMapping">
        <atom:title>SelfReportSkillMapping</atom:title>
     </collection>
     ▼<collection href="SpotAwardProgramAdvancedSettings">
        <atom:title>SpotAwardProgramAdvancedSettings</atom:title>
     </collection>
     ▼<collection href="TodoAction">
        <atom:title>TodoAction</atom:title>
     </collection>
     ▼<collection href="UpsertResult">
        <atom:title>UpsertResult</atom:title>
     </collection>
     ▼<collection href="InterviewOverallAssessment">
        <atom:title>InterviewOverallAssessment</atom:title>
     </collection>
```

Figure 1-6. *Results from a Simple ODATA Request*

What did we really just do? We executed a *GET* request, meaning we asked for data without requesting any changes to the SuccessFactors database. A POST request would be used to make such changes. We didn't specify a type of entity that we wanted to look at, such as requisition. Every ODATA request is done using an http(s) URI (Uniform Resource Identifier). We didn't specify any parameters to say which data we wanted to return, so the full data model was returned. A browser can be used exclusively to perform requests. Once you start working with different environments, string requests along so that request parameters are based on the results of previous requests, and other needs such as automated testing, you need to invest in an API app, as we'll see in Section 1.3. The small learning curve of becoming familiar with an app is far outweighed by the benefits.

Example: Reading a Requisition Using Postman

Our goal is to retrieve all requisitions that have a status of "Approved." We will read the *JobRequisition* entity, and filter on a field labeled "Internal Status." The field name is *internalStatus*. The value of this field is changed when requisitions are created, approved, and closed.

Getting Started with Postman

There are many desktop applications and browser-based tools for the testing and automation of interaction with HTTP APIs. We are going to use Postman as it has a free version and is one of the more commonly used ones. It's also great for collaboration. You may be working with a colleague such as an integration consultant or back-end web developer. Being able to quickly build API calls and export them for use by your colleague can be a great productivity boost.

The Postman app can be found here www.postman.com/downloads/. It is available for Mac, Windows, and Linux. There is also a web version of Postman. We shall demonstrate using the desktop app. We will teach just enough Postman to perform the necessary API calls but recommend you check out the Postman Learning Center at https://learning. postman.com/docs/getting-started/introduction/. Figure 1-7 shows the initial view of Postman.

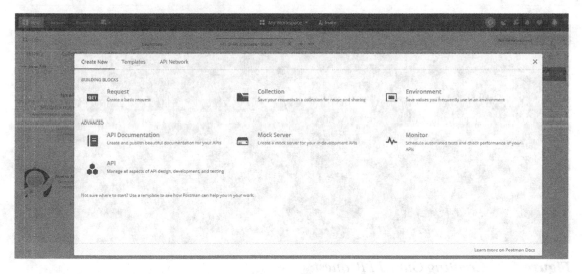

Figure 1-7. *Initial View of Postman or by Clicking New*

We are going to start by creating a New Environment. This means we won't have to remember the SuccessFactors API URI. In the future, it will also make it simple to switch between demo or customer environments.

Click "Environment" and then change the name from New Environment to something meaningful like "Mark's Demo Environment." Below that, add a variable name of *Host* and an initial value of the ODATA API URL for your datacenter, such as https://apisalesdemo8. successfactors.com/odata/v2. Add a variable for *username*, making it user@companyid.

Create a variable called *password* and set its value. Note that these examples use basic authentication for demonstration purposes - certificates will soon replace basic authentication OData calls by November of 2022. Make sure the environment name next to the eye on the right side is your environment name instead of "No Environment."

Executing a Request Using Postman

Now that we have Postman installed and our environment set up, we can create our first request. Click the + button near the top left of the screen to create a new request.

For the method to the left, make sure that GET is selected. For the host, enter *{{Host}}/JobRequisition*. This will use the correct host and request JobRequisition entities.

Under the Params tab, enter a KEY of "internalStatus" and a VALUE of "Approved." Note that the parameters will automatically update the URL. You may also enter a description. In Figure 1-8, we use the description to include the valid job requisition status options.

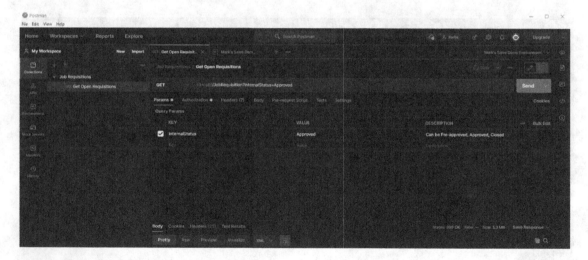

Figure 1-8. *Creating Our API Request*

Click the "Authorization" tab and select "Basic Auth." Oauth 2.0 should be used in the real world. We are using Basic Authentication for convenience. For username, enter {{Username}}, and for password, click the eye to display the password and type {{Password}}. Authentication values will be automatically passed from your environment. One advantage of using these variables is that you can share your API requests without sharing login information.

Figure 1-9 shows the authorization options for the service within Postman.

Figure 1-9. *Basic Authorization Information*

Save your API request by clicking the save disk icon to the right. Give it the name "Get Open Requisitions." Click "+ Create Collection" and call the collection "Job Requisitions."

Figure 1-10. *Saving Request and Creating a Collection*

End Result

Click Send to execute the request. You will see results as they appear in Figure 1-11.

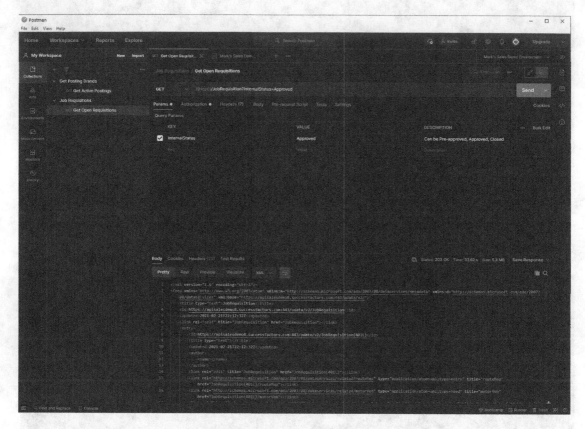

Figure 1-11. *Results from Request for Job Requisitions*

Status 200 in green to the right indicates that our request was successful.

Each requisition returned is represented by an <entry> element. If you scroll down, you will see the field values within the requisition, as in Figure 1-12.

Figure 1-12. *Field Values of Job Requisition 401*

If you want to save the results offline to share or review later, click "Save Response" on the right side.

Conclusion

You have taken the first steps in unleashing the power of the SuccessFactors ODATA API. You now have an understanding of how to use Postman to make a call to retrieve job requisitions. You've learned how to set up a Postman environment to easily switch which SuccessFactors instance to call. In addition, you know how to use parameters to filter job requisitions.

Whether you are a SuccessFactors system administrator or not, you have learned the necessary permissions for making ODATA calls. In addition to experimenting in your own test instance, this will allow you to advise system administrators on necessary permissions and to perform troubleshooting.

CHAPTER 2

Requisition Create and Update Automation

In Chapter 1, we walked you through the basics of making ODATA calls. In this chapter, we walk you through the first practical application of making those calls in the recruiting process: requisition creation. We start by walking you through some practical business scenarios that might trigger the need to build such an automation. Then we review the relevant SAP SuccessFactors ODATA API objects you would use to construct such an automation. Last, we put all you've learned thus far to walk you through two case studies. Let's get started!

Business Scenarios

In this section, we walk through some practical business scenarios to automate requisition creation in SAP SuccessFactors Recruiting.

To get started, it is important to know that if you are an Employee Central or SAP on-premise customer, and want to integrate and automate requisition creation from these systems into SAP SuccessFactors Recruiting, SAP offers standard integrations and automations between these systems. We will therefore not cover these scenarios, but rather scenarios where third-party systems are involved or automation outside of the standard is needed.

Simple Scenario: Small Business Payroll

In our simplest scenario, a small business of 2,500 employees has grown rapidly and plans to grow even further. To aid this growth, they would like to implement SAP SuccessFactors Recruiting. The company only has a payroll system and isn't ready yet for a full Human Resources Information Management System (HRIS) with position

15

© Anand 'Andy' Athanur, Mark Ingram and Michael A. Wellens 2022
A. A. Athanur et al., *Innovative SAP SuccessFactors Recruiting*, https://doi.org/10.1007/978-1-4842-7425-5_2

management capability. However, they do have a few key position types that are hard to keep filled. Thus, they would like a trigger to update the number of vacant positions on a few key requisitions as people are hired/terminated in the payroll system. These few requisition IDs could be set up in a config table or as simple variables in the source payroll system, so we avoid having to establish two-way communication and can just send simple updates to these preexisting requisitions. This would keep development and implementation to a minimum. This scenario is illustrated in Figure 2-1.

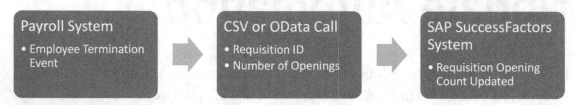

Figure 2-1. *Single Payroll System Updates Requisition Open Position Count with One-Way Communication*

Note   *Many payroll systems may not be robust enough to make ODATA calls on their own and may need middleware to negotiate the calls. File-based transmission may also be a valid option to consider if real-time transmission is not required.*

Mid-complexity Scenario: Mid-size Business HRIS

In a second scenario, a mid-size business of 10,000 employees has a combined HRIS and payroll system with position management capabilities. They have decided to implement SAP SuccessFactors Recruiting and want to keep tight integration between the vacant positions in the HRIS system and open requisitions in SuccessFactors Recruiting. Thus, any time a position object is empty and not closed in the HRIS, a requisition is triggered in SuccessFactors. If the position is closed on the HRIS, the requisition is closed in SuccessFactors Recruiting. Thus, two-way communication is required to synchronize the positions with the requisitions – for example, by sending back the requisition ID. This way, updates and closures could also be tracked on one or both systems and keep the position and requisition objects synchronized. Figure 2-2 illustrates this scenario.

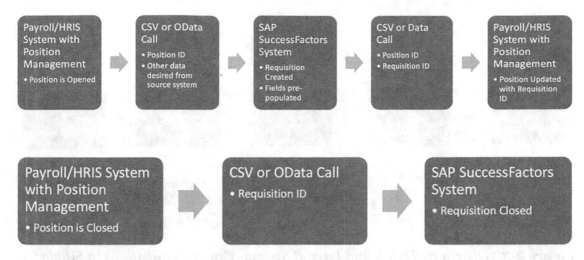

Figure 2-2. *Single HRIS/Payroll System Triggers Requisition Creation with Two-Way Communication*

Complex Landscape Scenario

In a third scenario, a holding company of 80,000 employees has made several acquisitions of smaller businesses around the world that each have their own HRIS and/or in-house/third-party payroll systems within each disparate country but no recruiting system. These systems maintain anywhere from 10,000 to 20 employees across 25 countries. The business has made a case that it does not make sense to immediately consolidate to a single combined HRIS and payroll system such as Employee Central because this would require significant up-front capital with little benefit that is visible to employees or potential employees. However, in this particular business, talent is hard to find. Furthermore, the parent company has a strong brand, whereas the held companies are virtually unknown and often compete for resource against one another. Thus, it makes sense for the company to have a single point of entry and repository for recruiting candidates to help coordinate effort and take advantage of the strongest branding. This would also allow the company to build consolidated talent pools in their highly competitive business amidst the "war for talent." It would also offer the company a best time to value and strategic competitive advantage for their particular business. In this scenario, it makes sense for multiple different HRIS systems to create requisitions in a single SuccessFactors Recruiting system and also allow manual requisition creation in lieu of integration for some of the smallest countries. Figure 2-3 illustrates the technical architecture of this scenario.

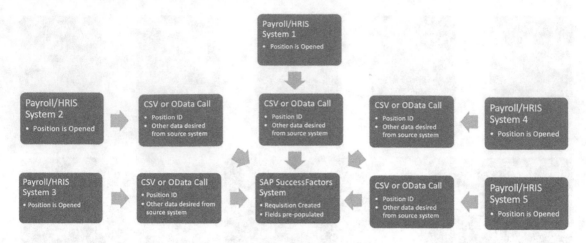

Figure 2-3. *Distributed HRIS and Payroll Systems Create Requisitions in Single SAP SuccessFactors Recruiting System*

Relevant ODATA API Objects

Before we dive into creating a solution in our case study, it is essential to know what tools and data are available to support these scenarios we have described.

The Job Requisition Object

The main object to be manipulated in these scenarios would of course be the requisition object. Like any configurable object based on a template, the requisition object will have baseline fields common to any requisition in any system and will also have fields specific to each template. The template configuration will not only determine what fields exist but also which are required and what permissions exist on each.

Checking the Metadata Structure of the Job Requisition Object

Since requisition fields will always vary from implementation to implementation, we should always check the metadata structure of the requisition in the system before getting started. We assume that the system where the interface is to be created has already been configured to the customer's specific recruiting requirements (i.e., at least one requisition template and other recruiting templates have been configured). There are a couple of ways to view the structure of the requisition object. First, the first method would be to look in the ODATA library. To do this, follow the given steps (be sure you are assigned the "Integration Center" permission first as well as all recruiting permissions associated with reading and creating ODATA objects:

1. Type and select "Integration Center" in the search bar.

2. The main menu of the Integration Center will appear. Click "Data Model Navigator."

3. In the new pop-up, type "JobRequisition" in the search and check the box next to "JobRequisition" and click "close."

4. The entity will appear as shown in Figure 2-4.

```
                                                                      JobRequisition CRUDU
♀ Job Requisition Id (jobReqId)
  Age (age)
  Application Template Id (appTemplateId)
  Assess Rating Scale Name (assessRatingScaleName)
  Candidate Progress Status (candidateProgress)
  Date Closed (closedDateTime)
  Requisition Comments,Comments,Comment (comment)
  Commission (commission)
  Corporate Posting (corporatePosting)
  CostCenter Code (costCenterCode)
  Cost of Hire (costOfHire)
  Country (country)
  Date Created (createdDateTime)
  Currency (currency)
  Job Level (customString3)
  Job Status (customString4)
  Job Pay Grade (customString8)
  Default Language (defaultLanguage)
  Is Deleted (deleted)
  Department Code (departmentCode)
  Division Code (divisionCode)
  ERP Amount (erpAmount)
  Evergreen (evergreen)
  Facility (facility)
  Form Data Id (formDataId)
  Form Due Date (formDueDate)
  Hiring Manager Note (hiringManagerNote)
    QUESTIONNAIRE  (instrCompQuest)
    POSITION DETAILS  (instrContractDetails)
    Cost of Hire Report Data  (instrCostHire)
```

Figure 2-4. Example View of JobRequisition Entity in the Data Model Navigator

---

Note   *You can choose multiple entities in this pop-up to visualize the relationships between them.*

In our experience, the Data Model Navigator layout is difficult to read. Sometimes if you simply want to look at what fields exist in the entity, you can follow the given steps:

1. Type and select "Integration Center" in the search bar.

2. The Integration Center main menu will appear. Click "My Integrations."

3. Click the "Create" button.

4. In the drop-down that appears, select "Scheduled Simple File Output Integration."

5. In the "Search for Entities by Entity Name" field, enter "JobRequisition."

6. Choose the first option that appears below "Job Requisition."

7. The screen will update on the right with a listing of all of the fields and navigation within the JobRequisition Entity. An example is shown in Figure 2-5.

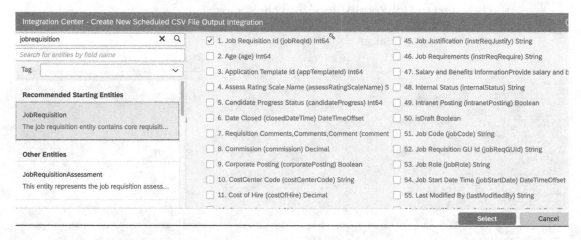

Figure 2-5. *Viewing Job Requisition Fields from the "Create New Scheduled CSV File Output Integration" Screen*

Here, you are able to examine what fields exist. There are a few field naming conventions and concepts to note here:

1. The text label of the field is shown first, the values in parenthesis are the technical field names, followed by the data type.

2. The Job Requisition Id is the primary key (as indicated by the key icon).

3. Fields beginning with "instr" are read-only visual labels/ instructions for the user that contain no values – you can typically ignore these fields for the sake of interfaces.

4. Typically, SAP and implementation consultants know to start custom fields with "cust" – however, there is no hard system requirement to do this.

5. Navigations are fields that link the job requisition to other entities by the primary key of that related entity. For example, any picklist field would contain the PicklistOption ID of the value for that particular requisition, or "Last Modified By" would contain the user ID of the user who last modified the requisition which could be used to look up more details about the user.

6. The fields shown are an amalgamation of all fields across all requisition templates loaded in the system. It is important to select the proper template ID when creating a requisition so that you create valid fields for the specific template.

Exporting Sample Requisition Data

Before attempting to create an interface that will create a requisition, it is a good idea to export requisition information from manually created requisitions. This way, you can obtain solid examples of how each field value should be formatted from data that has already been confirmed valid by the system checks used to create a requisition specific to your system configuration. Taking this advice, let's continue where we left off in the system looking at fields to create a sample .CSV file with valid field values. Follow the given steps from the same screen:

1. Select the fields and navigations you wish to output in the file by clicking the checkbox next to each. At a minimum, we would recommend selecting all of the required fields per your specific template ID as well as any other fields desired for the interface.

2. When you are finished, click the "Select" button.

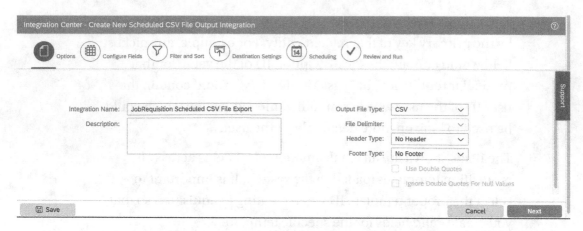

Figure 2-6. *Integration Center Scheduled CSV File Output Creation Screen*

3. You will be taken to the screen as shown in Figure 2-6. Give the integration a name and be sure to select Output File Type as "CSV" and Header Type as "Simple Header." Click "Next."

4. You will be taken to a screen showing sample values from your system for the fields you selected in the prior screen. An example is shown in Figure 2-7. Review the fields and click "Next."

Note *Some fields will not show true to their real-value formats on this preview screen! It is best practice to actually generate and download the .CSV file to see the true format. This will require you to have an SFTP server available. SAP SuccessFactors typically provides an SFTP server user ID / password to all customers.*

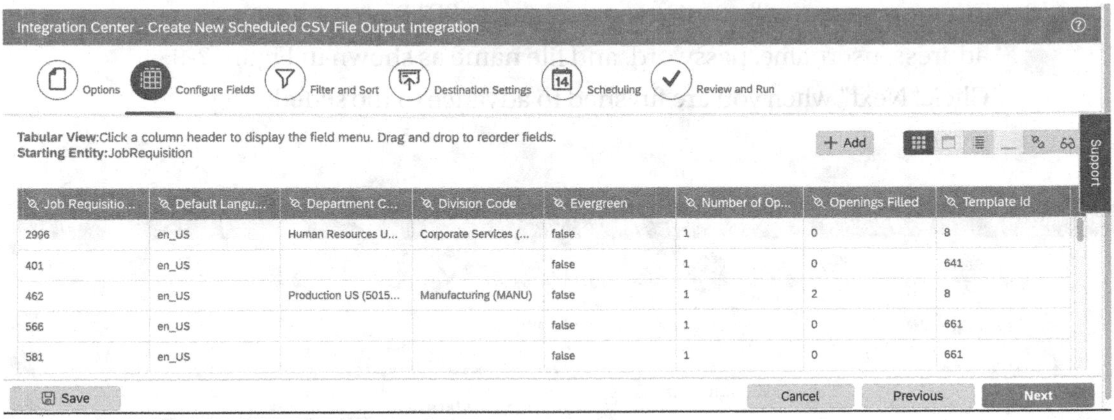

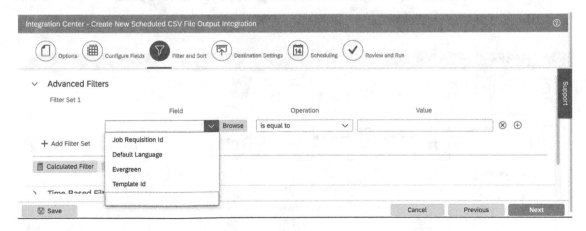

Figure 2-7. *Example Field Values Screen*

5. The next screen, shown in Figure 2-8, will allow you to filter
    your results. You can use this screen to find the particular job
    requisition(s) you manually created as an example of what
    the interface would create. You can click the caret icon next to
    "Advanced Filters" and then select the Job Requisition Id field to
    filter in this manner. You can also go back to the previous screen
    after setting the filter to check that you are getting the desired
    results (the screen will update to reflect only those values that
    match your filter criteria). When you are finished, click "Next."

Figure 2-8. *Selecting Filter Criteria*

6. The next screen will require you to enter an SFTP server host address, username, password, and file name as shown in Figure 2-9. Click "Next" when you are finished to advance to the scheduling screen.

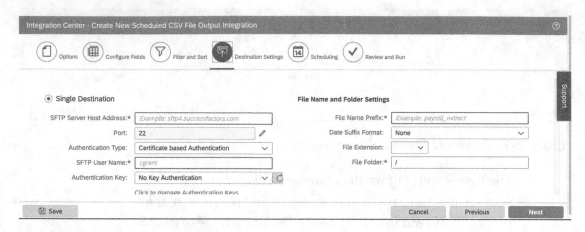

Figure 2-9. *SFTP Settings Screen*

7. You can skip the scheduling screen since this will be a one-time run. Click "Next" on this screen. Then click "Run Now" on the Review and Run screen as shown in Figure 2-10. You can check on the status of the run by clicking the refresh button next to "Last Run Time." (The system may also prompt you to save your integration first if you have not done so yet).

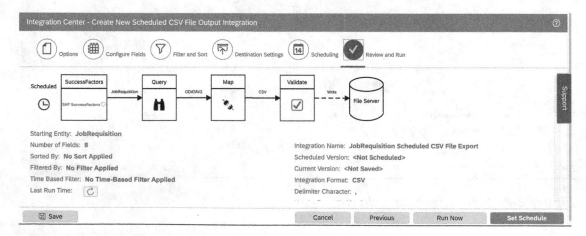

Figure 2-10. *Review and Run Screen*

8. Once the run finishes, log onto the SFTP server and download the file. You can open it up to view in a text editor as shown in the example in Figure 2-11.

```
Job Requisition Id,Default Language,Department Code,Division Code,Evergreen,Number of Opening,Openings Filled,Template Id,Job Requisition Id-JobRequisitionOperator,Job Requisition Id-J
2586,de_DE,Maintenance (DE) (5010134),Manufacturing (MANU),false,1,0,1440,2586,0
2570,en_US,,,false,1,0,1141,2570,0
2568,en_US,,,false,1,0,1141,2568,0
153,en_US,,,false,1,0,8,153,0
2575,en_US,,,false,1,0,1141,2575,0
155,en_US,,,false,1,0,8,155,0
2573,en_US,,,false,1,0,1141,2573,0
1762,ru_RU,{},Development (DEV),false,1,0,8,1762,0
1761,ru_RU,{},Development (DEV),false,1,0,8,1761,0
160,en_US,,,false,1,0,8,160,0
149,en_US,Planning & Scheduling FR (50130014),Manufacturing (MANU),false,1,1,8,149,0
163,en_US,,,false,1,0,8,163,0
2585,en_US,,,false,1,0,1440,2585,0
1402,en_US,,,false,1,0,901,1402,0
1244,en_US,Information Technology BRH (50171015),Information Technology (IT),false,1,0,8,1244,0
161,en_US,,,false,1,0,8,161,0
1243,en_US,Information Technology BRH (50171015),Information Technology (IT),false,1,0,8,1243,0
```

Figure 2-11. *Example .CSV File Output for Selected JobRequisition Object Fields*

Note *In our experience, programs like Microsoft Excel will reformat fields like dates. It is best to view files in a text editor such as Notepad++ or BBEdit to see the proper formatting of each field.*

Congratulations! You've built a simple .CSV file output! In addition, it is also a solid example of what an input file would look like to create or update a requisition. In the next section, we will take the example further in a case study where we create a working requisition update interface.

Case Study: Requisition Update Using CSV File

Now that we understand some practical business scenarios around requisition creation and update and we are familiar with the technical object available to manipulate in SuccessFactors, we can work on a practical application. In our first case study, we will look at our example first business scenario of the 2,500-employee company with a payroll system looking to update some preconfigured requisitions.

Simple Business Scenario Review

To quickly review, the example company has only a payroll system and no full HRIS system. When an employee that is in a mass hiring position is terminated from payroll, the payroll system will need to update the count of openings needed on a specific requisition(s) in the SAP SuccessFactors Recruiting system. Since many payroll systems cannot make ODATA calls on their own, it makes sense to use a nightly .CSV file transfer to an SFTP server where SAP SuccessFactors can pick up the file as the method of communication.

Implementation Steps

To get started, we can edit the file we created in the previous section. Follow these steps to prepare the file to be used as a template for the interface:

1. Open the file in your file editor.

2. Remove all rows except those for the requisitions you wish to update (e.g., your test requisitions).

3. Remove all column headers and associated row values except for the "Requisition Id" and the "Number of Opening" fields. An example is shown in Figure 2-12.

4. Save the file.

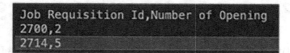

Figure 2-12. *Example Edited .CSV File Template*

Note   *Thinking about the interface, since the communication is one way, the payroll system will not know how many openings already exist in the SAP SuccessFactors Recruiting system. For this reason, we would normally use the Number of Opening column to indicate how many more openings to add based on the total number of terminations in the payroll system from key mass hiring positions. However, for the sake of simplicity in our example, we will just overwrite* *the value with whatever is in the file. We would need a middleware application to query the system for the current value, perform the calculation, and then update the field. You can learn more about middleware in the next chapter. If you have properly set up the configuration of your SAP SuccessFactors Recruiting system, the number of openings standard field count will be reduced by one automatically each time a candidate is put in the "Hired" status. Therefore, we will never need to send a negative/subtracting count from the payroll system on hire.*

Now that we've prepared our file, we have established what the file will look like coming out of the payroll system. We can hand it off to the payroll system developers to use as a template for what to output. On our end, let's use it to build our inbound interface into SAP SuccessFactors! Follow the steps to build the inbound interface in SAP SuccessFactors:

1. Type and select "Integration Center" in the search bar.

2. The Integration Center main menu will appear. Click "My Integrations."

3. Click the "Create" button.

4. Choose "Scheduled CSV Input Integration" as shown in Figure 2-13.

Figure 2-13. *Create Scheduled CSV Input Integration*

5. In the "Search for Entities by Entity Name" field, type "JobRequisition."

6. Click the JobRequisition object that will appear as the first search result on the left as seen in Figure 2-14. Then click "Select."

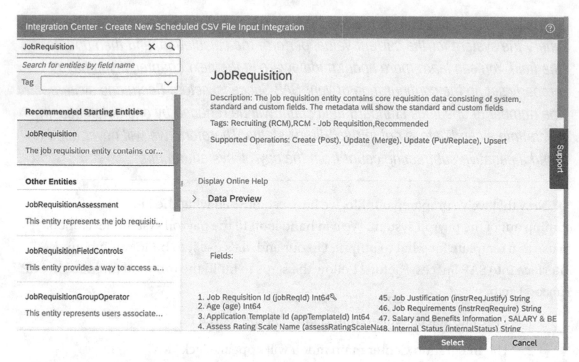

Figure 2-14. *Selecting JobRequisition Entity for the Inbound Interface*

7. Enter a name for the integration and then click "Next."

8. On the "Configure Fields" screen, click the "Upload Sample CSV" button in the upper-right-hand corner. Click "Browse." Choose your template you updated earlier in this section and then click "Upload." An example is shown in Figure 2-15.

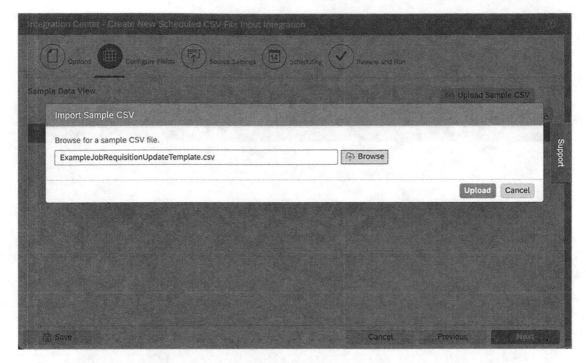

Figure 2-15. *Uploading the Integration File Template*

9. You will see the system update with the contents of your file as
    shown in Figure 2-16.

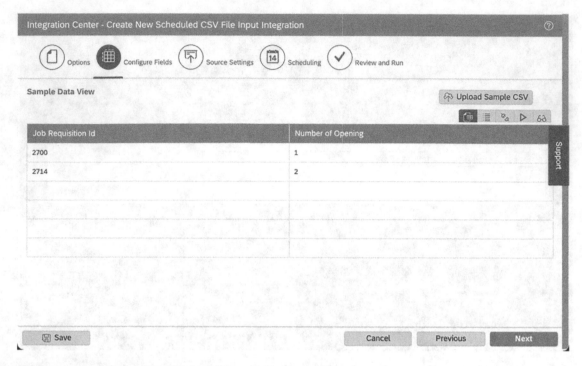

Figure 2-16. *Result of Uploading .CSV Template*

10. Now we will need to map each column in the file with the correct field in the JobRequisition object in SAP SuccessFactors. Click the icon on the right-hand side of the screen showing two plugs connecting. The screen will update to the Field Mapping View.

11. Ensure the "Operation" drop-down in the center is set to "Upsert Multiple." Drag and drop the source fields from your .CSV file template on the left to their corresponding JobRequisition object fields in the center. You will notice the icon in front of each mapped field in the center will change to closed plugs as each mapping is completed. Additionally, the sample ODATA code in the preview on the right will update as well (behind the scenes, the Integration Center is really just a front end that designs and runs ODATA calls). An example is shown in Figure 2-17.

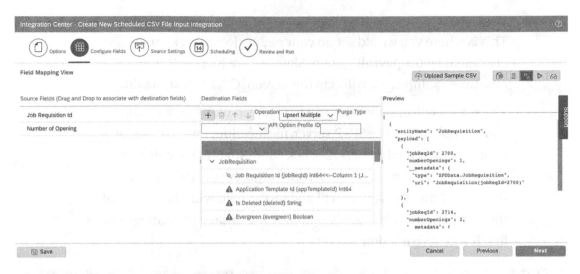

Figure 2-17. *Example Field Mapping View*

12. Click the "Next" button. The screen will update as shown in Figure 2-18. Enter your SFTP connection details and file naming convention.

Note *For a nightly batch update, we would recommend including the current date at the end of the file name. This is where you would use the "Date Suffix Format" field to define the format of the date that would appear at the end of the file name. You should also choose "CSV" for the "File Extension" field.*

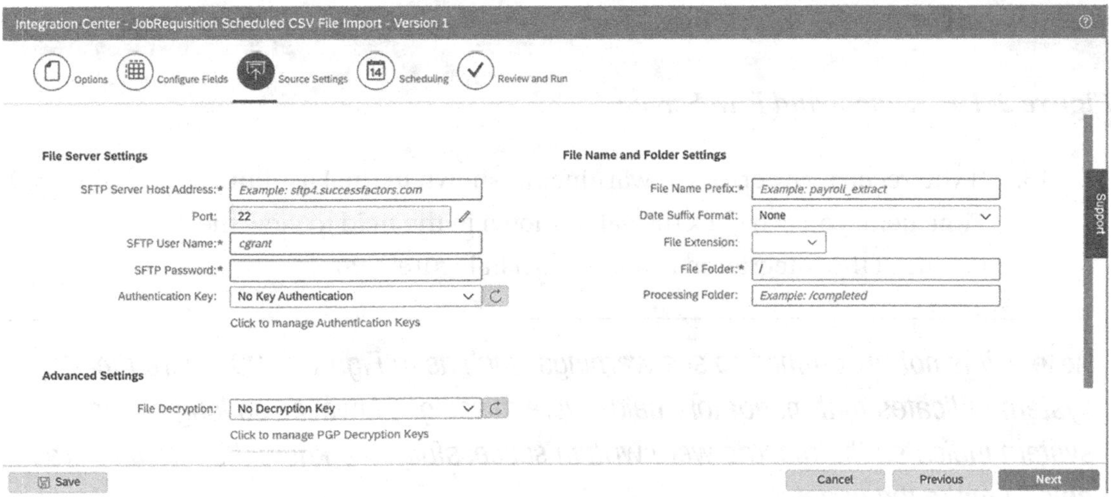

Figure 2-18. *SFTP Settings for Inbound Integration*

13.   Click "Next". The screen will update to the scheduling screen. This is where you would set up your nightly job to pick up the file generated by the payroll system. Since we are just performing a one-time example, we will skip this screen. Click "Next" again.

14.   The screen will update as shown in Figure 2-19. Be sure to upload your sample file to the SFTP server, file location, and use the file name you specified in step 12. Click "Run Now" to execute the integration. You can click the refresh button to the right of "Last Run Time," and the system will update the field with the results of the run. (The system may also prompt you to save your integration first if you have not done so yet).

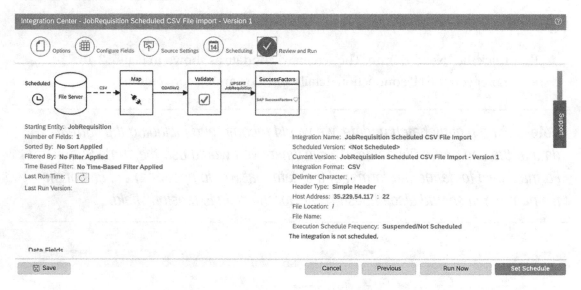

Figure 2-19.   Review and Run Screen

15.   If you receive any errors or warnings as shown in the Last Run Time field, you can click the value shown in the field to view the details of the integration run as shown in Figure 2-20.

Note   *It is not uncommon to see warnings, such as in Figure 2-20 where the system indicates that mandatory fields were missing. However, so long as the system indicates the records were written successfully, we know the interface was able to make the updates.*

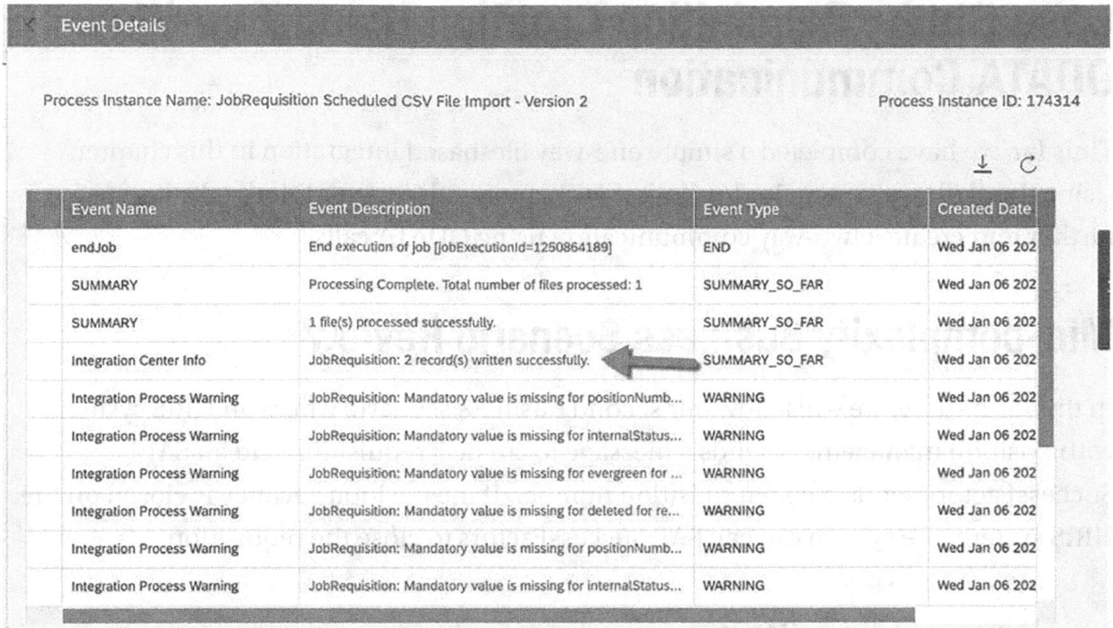

Event Name	Event Description	Event Type	Created Date
endJob	End execution of job [jobExecutionId=1250864189]	END	Wed Jan 06 202
SUMMARY	Processing Complete. Total number of files processed: 1	SUMMARY_SO_FAR	Wed Jan 06 202
SUMMARY	1 file(s) processed successfully.	SUMMARY_SO_FAR	Wed Jan 06 202
Integration Center Info	JobRequisition: 2 record(s) written successfully.	SUMMARY_SO_FAR	Wed Jan 06 202
Integration Process Warning	JobRequisition: Mandatory value is missing for positionNumb...	WARNING	Wed Jan 06 202
Integration Process Warning	JobRequisition: Mandatory value is missing for internalStatus...	WARNING	Wed Jan 06 202
Integration Process Warning	JobRequisition: Mandatory value is missing for evergreen for ...	WARNING	Wed Jan 06 202
Integration Process Warning	JobRequisition: Mandatory value is missing for deleted for re...	WARNING	Wed Jan 06 202
Integration Process Warning	JobRequisition: Mandatory value is missing for positionNumb...	WARNING	Wed Jan 06 202
Integration Process Warning	JobRequisition: Mandatory value is missing for internalStatus...	WARNING	Wed Jan 06 202

Process Instance Name: JobRequisition Scheduled CSV File Import - Version 2 Process Instance ID: 174314

Figure 2-20. *Detail of Last Interface Run Showing Records Written Successfully*

16. To double-check your work, you can also navigate to your requisitions and click the "i" icon in the upper-right-hand corner on the requisition details view. You should see the Change History show an ODATA update as seen in Figure 2-21. This is your integration interface making the update!

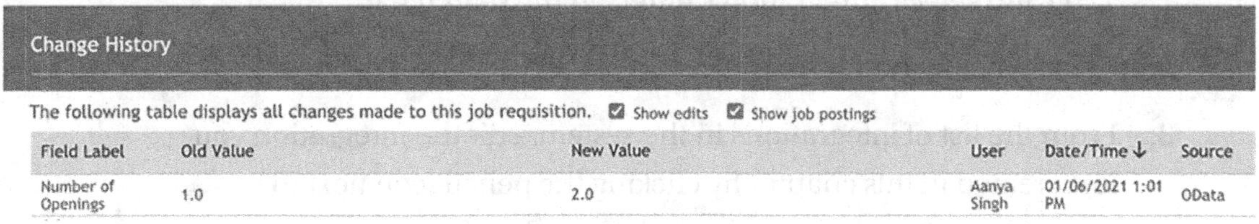

Change History

The following table displays all changes made to this job requisition.   ☑ Show edits   ☑ Show job postings

Field Label	Old Value	New Value	User	Date/Time ↓	Source
Number of Openings	1.0	2.0	Aanya Singh	01/06/2021 1:01 PM	OData

Figure 2-21. *Example Change History for Requisition Updated by Interface*

You've done it! You have now completed a simple and practical requisition update interface! For our next case study, let's get a little more complex and directly use ODATA rather than using the Integration Center. We can also establish a two-way connection rather than just one way.

Case Study: Requisition Creation Using Two-Way ODATA Communication

Thus far, we have completed a simple one-way file-based integration in this chapter. Using the skills we've acquired in both this chapter and Chapter 1, let's take it a step further and create a two-way communication using ODATA calls.

Mid-complexity Business Scenario Review

In this case study, we will follow our second business scenario where an HRIS system with position management sends a message to create a requisition and an SAP SuccessFactors sends back a requisition number. If the position vacancy is closed on the HRIS system, the system can call SAP SuccessFactors to close the requisition.

Implementation Steps

For the outgoing call to create the requisition, we can simply modify the integration we have already created to gather an example ODATA call that will be used in lieu of a .CSV file row! You will recall in Figure 2-17, we saw the ODATA call that SAP SuccessFactors Integration Center makes when reading the .CSV file. Similarly, we can create another integration and use the example call it creates as the template ODATA call that is made from the HRIS system (or middleware) to create the requisition. Follow the steps to create the integration and create the example ODATA call:

1. Type and select "Integration Center" in the search bar.

2. Click "My Integrations."

3. From the list of integrations in the system, edit the integration you last created in this chapter by clicking the pencil icon next to its name.

4. Click "Configure Fields" at the top and then click the plug icon in the upper-right-hand corner. The screen should appear like Figure 2-17.

5. Under "Operation," change the selection to "Create/Post."

6. You will notice multiple fields will move to the top and have an exclamation mark to the left of them. These are required fields you will need to provide in order to create a requisition successfully. An example is shown in Figure 2-22. An explanation of each field is listed in Table 2-1.

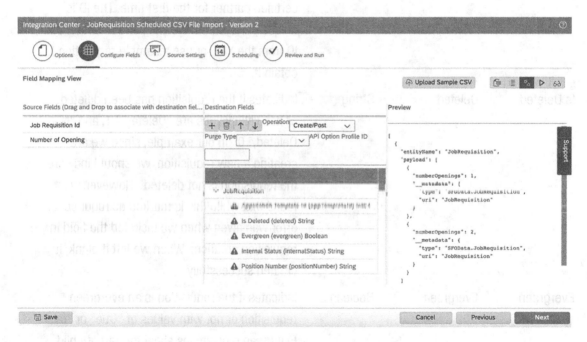

Figure 2-22. *Example of Fields Required to Create a Requisition*

Table 2-1. *Required Fields to Create a Requisition*

Field Name	Technical Name	Field Type	Description
Application Template Id	appTemplateId	Int64	Indicates which XML template to use to create an application when someone applies to the requisition (which defines the fields and permissions of the application). This corresponds to the unique identifier assigned to an application template when the template XML is uploaded to provisioning by SAP or a SuccessFactors-certified partner for the first time. The ID is unique per system. We will need to obtain this ID from the XML or use a sample requisition to obtain it.
Is Deleted	Deleted	String	Indicates if the requisition has been deleted or not. Valid values are "Deleted," 1, and "Not Deleted," 0. In our example, since we are creating a new requisition, we should indicate the requisition is not deleted. However, we found that while this is marked as required, an error was given when we included the field in requisition creation. When we left it blank, it created successfully.
Evergreen	Evergreen	Boolean	Indicates if the requisition is an evergreen requisition or not with values of "true" or "false." Evergreen requisitions allow for parent/child relationships and are used for high-volume positions where a one-to-one relationship is desired for each requisition and position. For our example, we will indicate the requisition is not an evergreen.

(continued)

Table 2-1. (*continued*)

Field Name	Technical Name	Field Type	Description
Internal Status	internalStatus	String	Indicates where the requisition is in the process, specifically as "Pre-approved," 0 (has not gone through the route map); "Approved," 1 (route map has been completed); or "Closed," 2. This defines which set of permissions will be used when viewing the requisition as an end user well. For our example, we will set the requisition to "Pre-approved" since it will be brand new and still have to go through the requisition approval process.
Position Number	positionNumber	String	If Employee Central is activated and integrated, this field is required and corresponds to the Employee Central Position number (it shows in our example since our demo system also has EC). Most likely, this field will not be used in our discussed business scenarios.
Application Status Set ID	statusSetId	Int64	Unique Identifier determined by the system when the Applicant Status Set Configuration is created the first time. This is unique by system as well. The Applicant Status Set defines what statuses will exist throughout the application process (such as applied, background check, hired, etc.). We would need to gather this information from the person who configured it, or for our example, we will grab this ID from a preexisting requisition.
Template Id	templateId	Int64	Indicates the XML template to use when creating the requisition. This XML defines the fields and permissions that exist in the requisition. We will need to get this ID from the XML or, as in our example, use an example of an existing requisition.

Note   *One of the most important decisions when setting up a requisition creation interface is determining who the originator will be. Since the first step after creating a requisition in SAP SuccessFactors is kicking off the route map, the originator needs to be determined since they are the one who will have access to the requisition at the first step in the route map. In our experience, the originator is often the primary recruiter. You will need to send the user ID of this originator in your interface any time you create a requisition. For both internal candidates who apply and recruiters or other requisition team members, you will need to make sure the user IDs exist for each of these in the SAP SuccessFactors system. This will require setting up a standard user data file integration or manual import. We will not cover the UDF in this book. For more information, see SAP note 2088000. For the sake of simplicity, we will leave the originator blank, and it will default to the user who logged on and created the requisition via the ODATA call.*

7. We will need to export an example requisition to gather what all of our fields should be. Click "Save" and then "Save As" to create a new import integration. We will return to it later. "Cancel" to close the new integration and return to the "My Integrations" screen.

8. Edit the example requisition export integration we first created in this chapter by clicking the pencil icon next to its name.

9. Click "Configure Fields" at the top and then click the "+ Add" button.

10. Add any of the required fields listed in Table 2-1 not already included in your output. In addition, add any fields that are required in the requisition XML. An example is shown in Figure 2-23. You may also want to include any other fields you wish to default from your HRIS system.

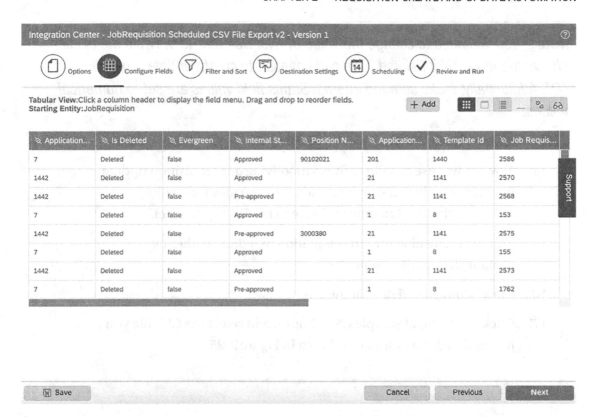

Figure 2-23. *Example Requisition Export with All Fields Required to Create a Requisition*

11. Click the "Save" button and then "Save As." In the pop-up, give the new export a name and click the "Save" button.

12. You may also want to give the file a new name on the "Destination Settings" screen. Click "Review and Run" when you are ready to run, and then click the "Run Now" button.

13. Open the file in your text editor, and choose a row with the desired values according to Table 2-1. Delete any unneeded rows that do not contain your desired sample data. An example is shown in Figure 2-24.

```
Application Template Id,Is Deleted,Evergreen,Internal Status,Position Number,Application Status Set Id,Template Id,Default Language,Country,id-PicklistOption,id-PicklistOption,id-PicklistOption,
7,0,FALSE,B,3000938,Y,8,en_US,United States,5616,11626,11962,3979,cbraun
```

Figure 2-24. *Example .CSV Used for Import to Create Requisition*

Note   *You will need to change the value of "Not Deleted" to 0. While the export shows the label, "Not Deleted," the code needed to create/update requisitions is "0". Additionally, make sure the Internal Status field values are set to numerical values.*

14. Now we will use this file to finish modifying our requisition update integration to become an integration that creates requisitions. Return to the "My Integrations" screen by clicking "Cancel."

15. Edit the requisition import integration we edited earlier by clicking the pencil icon next to it.

16. Click "Configure Fields" at the top.

17. Click the "Upload Sample CSV" button and select the CSV file you just updated. An example is shown in Figure 2-25.

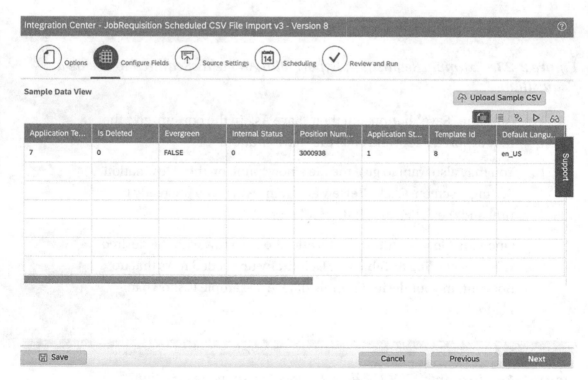

Figure 2-25. *Example Result of Uploading Requisition Creation CSV*

18. Click the plug icon to change to the field mapping view. Make sure the operation field is still set to "Create/Post." Drag and drop each field from the CSV file on the left to its associated field. You will notice the ODATA call example on the right updates as the field associations are made. An example is shown in Figure 2-26.

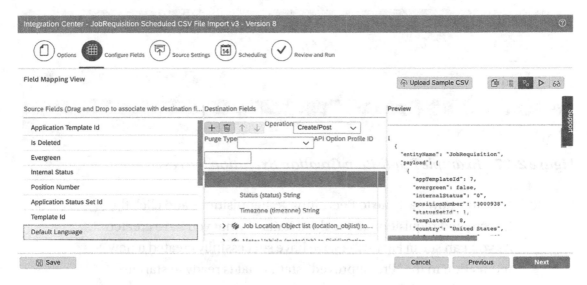

Figure 2-26. *Result of Mapping Fields*

19. You can now test the integration by clicking the play icon in the upper-right-hand corner of the screen and then clicking the "Run Preview Records" button. The system will either display a success message in the first column as shown in Figure 2-27 or show an error. If you receive errors, simply mouse over the "Run Result Status" column to view the error details, and execute again once you have addressed the issues.

Note   *Chances are you will run into an error on your first test execution. This is why we recommend using the Integration Center that has better access to data structures rather than blindly creating ODATA calls. The most common error is missing required fields. You may need to troubleshoot/find valid values in an example requisition and perform additional mappings until you receive a success* *message. This error is best fixed by finding example fields in an already existing requisition and creating mappings for those fields from your example export CSV that you have uploaded as the template for your requisition creation interface. You can always modify the interface to troubleshoot until you have resolved all errors.*

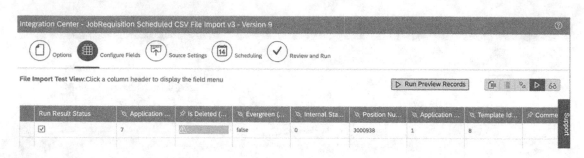

Figure 2-27. *Example Requisition Creation Execution Success*

20. You may now navigate to Recruiting ➤ Requisitions and click the last created requisition to see the details of what you have created. As you can see in Figure 2-28, we have successfully created a new requisition in the "Pre-approved" status that is ready to start its journey through the route map.

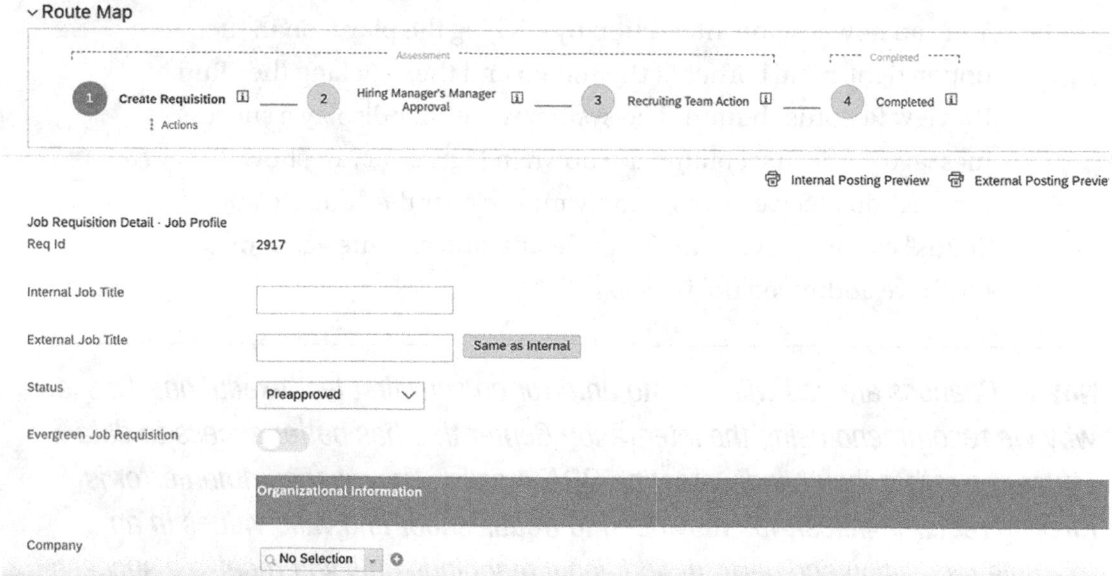

Figure 2-28. *Example End-User View of Successfully Created Requisition*

Congratulations! You have now created an example ODATA call; you can have your HRIS system (or middleware) populated with data to create a requisition! Our example is shown as follows. Using the skills you have acquired in this chapter and Chapter 1, you should be able to call the system to create a requisition!

```
[
  {
    "entityName": "JobRequisition",
    "payload": [
      {
        "appTemplateId": 7,
        "evergreen": false,
        "internalStatus": "O",
        "positionNumber": "3000938",
        "statusSetId": 1,
        "templateId": 8,
        "country": "United States",
        "defaultLanguage": "en_US",
        "filter1": [
          {
            "__metadata": {
              "type": "SFOData.PicklistOption",
              "uri": "PicklistOption(id=5616)"
            }
          }
        ],
        "filter2": [
          {
            "__metadata": {
              "type": "SFOData.PicklistOption",
              "uri": "PicklistOption(id=11962)"
            }
          }
        ],
```

```
      "filter3": [
        {
          "__metadata": {
            "type": "SFOData.PicklistOption",
            "uri": "PicklistOption(id=3979)"
          }
        }
      ],
      "hiringManager": [
        {
          "userName": "cbraun",
          "__metadata": {
            "type": "SFOData.JobRequisitionOperator",
            "uri": "JobRequisitionOperator"
          }
        }
      ],
      "location_obj": [
        {
          "__metadata": {
            "type": "SFOData.FOLocation",
            "uri": "FOLocation(externalCode='undefined',startDate=datetime
                '1900-01-01T05:00:00')"
          }
        }
      ],
      "state": [
        {
          "__metadata": {
            "type": "SFOData.PicklistOption",
            "uri": "PicklistOption(id=11626)"
          }
        }
      ],
      "__metadata": {
```

```
        "type": "SFOData.JobRequisition",
        "uri": "JobRequisition"
      }
    }
  ]
 }
]
```

Now that we have created the requisition, our remaining task is to make a subsequent ODATA call back to the HRIS system (or middleware) to indicate the requisition creation was successful and provide the requisition ID. To do this, we will simply create another ODATA query as we have done twice now. Our query will only need to include the HRIS system unique identifier of the position (such as position ID) and grab the unique ID of the requisition.

1. Type and select "Integration Center" in the search bar.

2. Click "My Integrations."

3. Click the pencil icon next to the last export integration you created to edit it.

4. Click "Configure Fields" at the top and then click the bulleted list icon in the upper-right-hand corner.

5. Remove all fields except the requisition ID and position ID.

6. Add the fields "jobReqId" and "positionNumber" (or whatever unique ID is used to store the HRIS system position unique ID) by clicking the "+ Add" button and searching for the fields.

7. Click "Filter and Sort" at the top and then click "Advanced Filters." Enter "Position Number" in the "Field" column and the number of the position in the "Value" column. An example is shown in Figure 2-29.

Note   *This assumes a 1:1 relationship between position and requisition. Also you may wish to use a custom field to store a third-party HRIS position number rather than the standard position field since this standard field is used by Employee Central.*

Figure 2-29. *Adding a Filter to the Query for the Specific HRIS Position Number*

8. Click "Configure Fields" to see the preview of the data. The query should return the requisition ID of the requisition you created in the last set of steps! An example is shown in Figure 2-30.

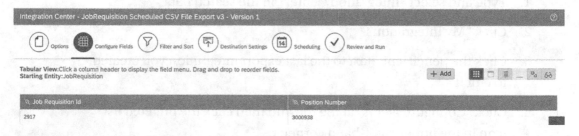

Figure 2-30. *Result of Query of Specific Position Number*

9. Click "Save" and then click "Export Integration Specification."

10. Open the specification with your text editor, and copy the query URL from the last line of the document.

Great! You now have an example query for the HRIS system to grab the requisition ID based on the HRIS system position number! You can now store this on the HRIS system and use it for subsequent calls such as to close the requisition when the position is closed. The example query is shown as follows:

```
/odata/v2/JobRequisition?$select=jobReqId,positionNumber&$filter=
positionNumber eq '3000938'
```

Using the skills you have now obtained, you should be able to work out how to close the requisition in the event the position is closed on the HRIS system (hint: save a new version of the requisition update integration we created in the first case study and update the internalStatus field to "2").

We have now successfully set up two-way communication between SAP SuccessFactors and an external HRIS system to open and close requisitions!

Conclusion

We hope you have enjoyed this step-by-step walk-through of creating and updating requisitions! By now, you should have a good understanding of some business scenarios that would call for creating and updating requisitions. In addition, you should be familiar with the requisition object and how to retrieve, update, and create data for this object. Furthermore, we've walked you through a couple of example case studies to show you examples of how to build some practical requisition update and creation integrations. In our next chapter, we will dive into further detail about creating requisitions using middleware.

CHAPTER 3

Requisition Automation Using Middleware

In Chapter 2, we walked through the automated creation and update of requisitions using Integration Center. Integration Center is great for simple use cases without complex logic. It's also a great tool for testing and troubleshooting ODATA calls before sharing with colleagues that are integrating from non-SAP SuccessFactors applications. When the requirement for updates goes beyond simple calculations, we need to look at other tools.

In this chapter, we will review commonly used tools for updating recruiting information and example use cases for each. We will then show two use cases: one using business rules and one using SAP Cloud Platform Integration (CPI) as a middleware integration tool. Job requisitions are used as an example in this chapter, but the same principles can be applied to data across all SAP SuccessFactors modules.

Middleware such as CPI is typically used to integrate with third-party solutions, for example, to integrate with an assessment vendor, a background-check solution, or a referral solution. However, sometimes CPI can also be used to automate the processing of data in SuccessFactors without any third-party integration. A common question is "Why are we using middleware when we don't have a third-party solution?" The answer is that it works. Limitations in SAP SuccessFactors business rules can often be overcome by the user of integration. Examples include the processing/creation of requisition questions, the automatic generation of offer letters, and many more. The reason to use CPI when integrating SAP SuccessFactors with SAP SuccessFactors is for data manipulation.

A. A. Athanur et al., *Innovative SAP SuccessFactors Recruiting*, https://doi.org/10.1007/978-1-4842-7425-5_3

Evaluating Solutions for Manipulating Recruiting Data

The SAP SuccessFactors HXM Suite is very powerful and flexible, with multiple built-in tools to manipulate data. Because the suite of modules is accessible using ODATA, it can also receive and send data to non-SuccessFactors applications. The four tools for data manipulation that we examine throughout this book are

- Integration Center

- Business rules triggered by the actions of the user (e.g., the recruiter)

- Business rules executed in batch

- SAP Cloud Platform Integration (CPI)

  Among the four solutions, we would only classify CPI as middleware, though Integration Center has some uses and functionality that also belong in middleware.

  So, what is middleware? Middleware is a solution that sits between two applications and turns the requests from one system into something consumable by another. Example tasks of middleware include restructuring the request format, mapping the field values from one system to those that are acceptable to another (example: one drop-down set of values to another), the use of logic to determine how to process a request, and error handling. Middleware also acts as a gatekeeper, to ensure that an application only recognizes a request from a particular IP address and therefore only handles valid requests.

  SAP Cloud Platform Integration is far from the only integration middleware provider. Other examples are Dell Boomi, MuleSoft, TIBCO, and Informatica.

There are three main factors to consider when choosing a tool to update requisition data:

1. **Business logic complexity** – Do you have If/Then/Else type logic such as "If Payscale Level > X, then Set Stock Option Eligibility to Yes"? Integration Center is limited in how complex this logic can be. Business rules should be the first tool to reach for. A great example of an Integration Center use case is if I wanted to update a field en masse due to a change in master data or picklist options. A file could be received with the new data and imported.

2. **Data to read and write** – The types of data available for evaluation and manipulation depend on the tool. Integration Center and CPI can read and write to anything in the SAP SuccessFactors data model. However, Integration Center is limited in that an integration can only deal with the reads/writes of one set of data at a time. For example, if I wanted to process some logic and then write to both an offer approval and a candidate application at the same time, CPI would be my only option.

   Business rules are very powerful but are limited in the types of data that they can read and write. Some examples of requisition information that are not consumable by business rules are requisition-specific application questions, metadata framework (MDF), and foundation objects that are associated with the requisition and not supported in the standard, such as position. As an example, if I wanted to pull address details from the location foundation object and use it to populate an email token, business rules would not support that.

   If the business logic used to manipulate data relies on storing the configuration in a custom MDF object, then CPI is the only option.

3. **Who is doing the work?** Business rules and Integration Center are designed to be used by a system administrator as the client. The client may have developers that can build integrations in CPI. If the work can be done in a simple way that doesn't need developers and the maintenance of a CPI integration, business rules should be the option used. If a batch business rule is needed, then provisioning access is required. SAP support or a certified consultant will be needed to execute and schedule the batch job.

The uses for each solution can be seen in Table 3-1.

Table 3-1. *Factors in Choosing a Solution*

Decision Factor	Batch Business Rules	User-Triggered Business Rules	Integration Center	CPI
Technical difficulty	Low	Low	Medium	High
Evaluation logic (If/Then)	Yes	Yes	Limited	Yes
Multiple objects to read/write	No	No	No	Yes
Access to all ODATA objects	No	No	Yes	Yes
Custom MDF Object-based business logic	No	No	No	Yes
Performed by client admin	With help	Yes	Yes	No

In looking at the table, it would appear that CPI is the best overall solution. However, CPI has some drawbacks and should be avoided if another solution exists:

- **Work effort** – CPI integrations typically involve a functional specification, a technical specification, extra development, and testing effort.

- **Technology debt** – One of the reasons that clients moved from on premise to the cloud is to stop owning custom solutions that need to be maintained over time.

- **Ownership and process agility** – If the recruiting system administrators can make changes as processes change, they can be more responsive to their client (the recruiter) needs.

This book will not get into the details of building a CPI integration. The intent is to teach you how to choose the right tool and, in the case of CPI, what requirements to share with your integration developer.

Case Study: Automatically Adding "Remote" to Remote Job Titles

A field to indicate a remote position has been added to the job requisition. The client wants job titles on the external facing career site as well as job boards to be the position name, followed by "(Remote)" if the remote hiring field is yes. The job title always defaults from the position and is not changeable by recruiters. The remote indicator defaults from the position but is changeable by the recruiter. For this reason, the external job title change needs to be automated.

The external job title is already populated using the business rule that is triggered when creating a job requisition from a position. We want to update the title when the value in the remote job field is changed by a user, such as a recruiter. We won't detail the steps to create a business rule here as this is very well documented by SAP.

We simply need to create a rule that is triggered by the onChange condition tied to the remote job field of the requisition. The rule then checks if Remote Job = Yes. If this condition is true, then the external job title is changed to the position name + "(Remote)". If it isn't true, then we set the job title to be the position name. This ensures the word "remote" is removed if the req is switched from a remote to non-remote hire.

Case Study: Automatically Adding Questions to a Requisition

Requisition Question Background

The job application entity in SAP SuccessFactors Recruiting is powerful and flexible, capable of capturing job application information or follow-up information requests based on the hiring country, status of the application (e.g., when the applicant first applies vs. when they accept an offer), and whether they are an existing employee or an external applicant and dependent on the requisition template used (e.g., professional hire vs. interns).

Questions are added to individual requisitions at the time of creation before they are posted to internal and external job boards. In Figure 3-1, we can see how questions are added to the requisition and in turn to the application form that applicants complete. Questions can be chosen from a predefined library or created at the time of the requisition creation. Though question selection is a simple process, selecting them is an extra couple of clicks, and more importantly, users forget to add screening questions which should be mandatory knockout questions. This can have a serious impact in the screening of applications.

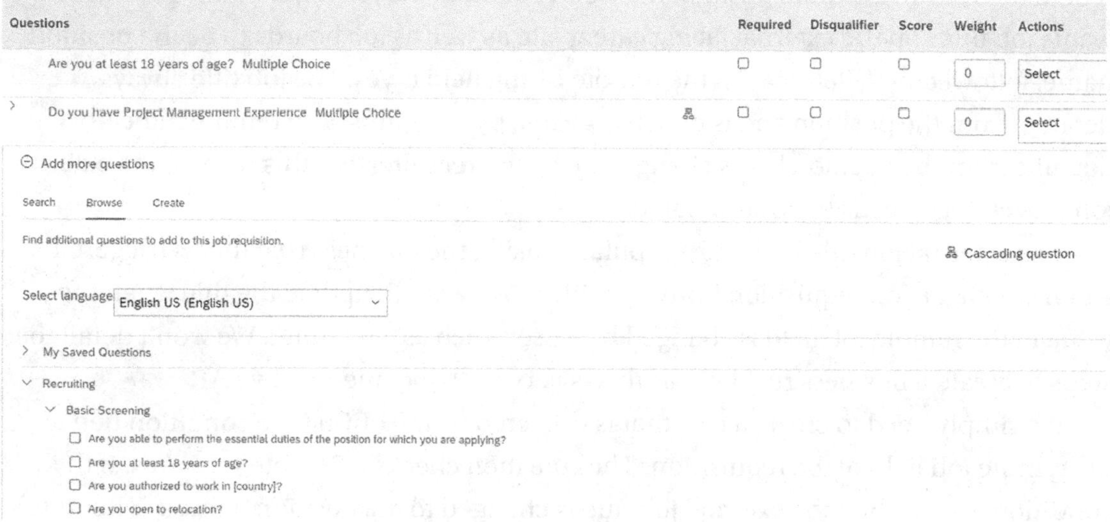

Figure 3-1. *Adding Questions to a Requisition*

Questions have the advantage of automatically disqualifying applicants based on answers, of having answers worth different scores, and having responses lead to other questions. The latter is called cascading questions. In Figure 3-2, we can see the addition of cascading questions to a requisition.

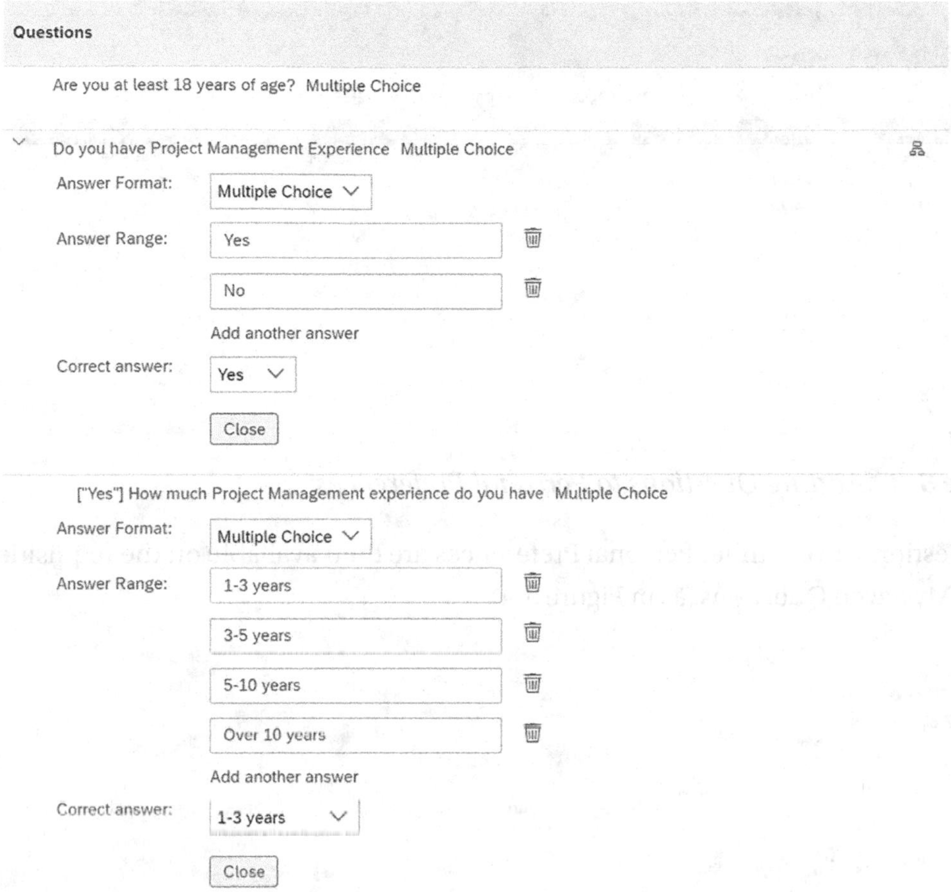

Questions

Are you at least 18 years of age?   Multiple Choice

⌄   Do you have Project Management Experience   Multiple Choice   ⚎

Answer Format: Multiple Choice ⌄

Answer Range: Yes   🗑

                    No   🗑

                    Add another answer

Correct answer: Yes   ⌄

                    Close

        ["Yes"] How much Project Management experience do you have   Multiple Choice

Answer Format: Multiple Choice ⌄

Answer Range: 1-3 years   🗑

                    3-5 years   🗑

                    5-10 years   🗑

                    Over 10 years   🗑

                    Add another answer

Correct answer: 1-3 years   ⌄

                    Close

Figure 3-2. *Adding Cascading Questions to a Requisition*

Maintaining a consistent library of prescreening questions organized by category helps to ensure consistent and unbiased screening of applicants. In the case that recruiters have specialized questions that are not in the library, they can be created under My Saved Questions, as in Figure 3-3.

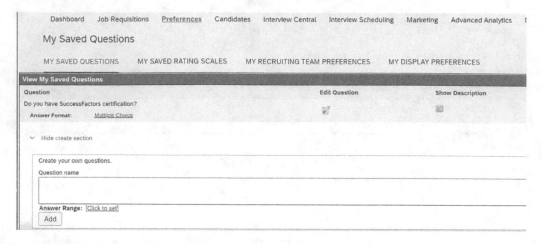

Figure 3-3. *Adding Questions to Personal Preferences*

Questions saved under Personal Preferences are then available on the requisition, under My Saved Questions, as in Figure 3-4.

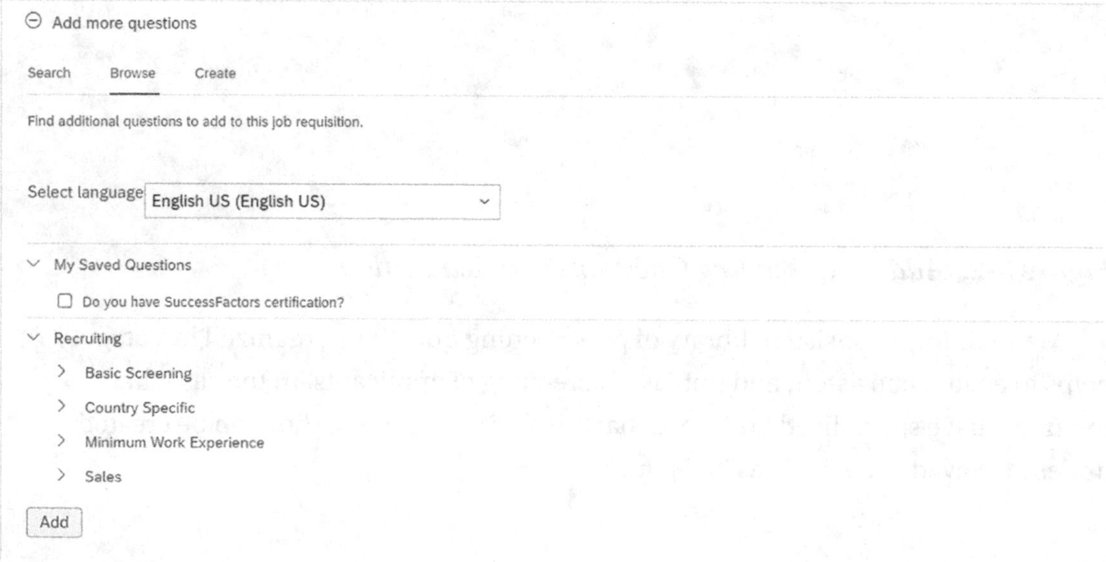

Figure 3-4. *Adding Questions to Requisition from My Saved Questions*

An organized library of application questions is the first step to ensuring consistent screening questions. However, there is still the human element of recruiters adding questions to the requisition based on the prescreening requirements. This is both additional effort in the creation of requisitions and introduces the possibility for human error. This is where automation comes in. Our options for defaulting requisition questions are limited. This type of data is not currently supported by business rules. This is where middleware comes in handy due to the access to all ODATA entity objects, such as job requisition screening questions.

For the purposes of this example, we are going to keep the automation simple and useful. For organizations that use application questions to automatically disqualify candidates, the requirement is to always include the correct disqualifying questions based on the hiring need. Recruiters are people too. Sometimes they forget to add necessary questions to a requisition. Worse still, sometimes questions are added after some applicants have completed their applications, leading to legal and compliance issues of treating applicants differently for the same job.

We will use an existing question from the question library and will use the requisition template to determine the question to add. Examples of requisition templates are interns, professional, country, or division specific. In general, the fewer requisition templates, the better for a homogeneous global business process and lower system maintenance effort. Other criteria to differentiate questions to be added could be job family, country, and many others.

Review of Relevant ODATA Entities

We can see from Figure 3-5 that multiple questions may be associated with a requisition and that a requisition question may have multiple choices. When multiple question choices are used for a question, they would not be associated with a text answer (type TEXT_QUESTION) or numeric answer (type NUMERIC_QUESTION).

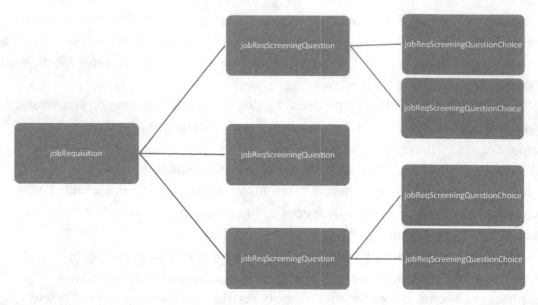

Figure 3-5. *Structure of Questions Associated to Job Requisitions*

You will also see the JobReqQuestion in the ODATA API Data Dictionary. This entity stores the attributes question category, question name, question source (which question library), and question type.

To better understand the relationship between the earlier entities, let's read some questions from a job requisition using Integration Center.

Requesting Job Requisition Question Choices Using Integration Center

To extract job requisition questions using Integration Center, create a new Scheduled CSV Output Integration using either JobRequisition or JobReqScreeningQuestion as the starting entity. In the example in Figure 3-6, we are starting with the JobReqScreeningQuestion. Typing Req or Question in the search bar will shorten the list of possible entities.

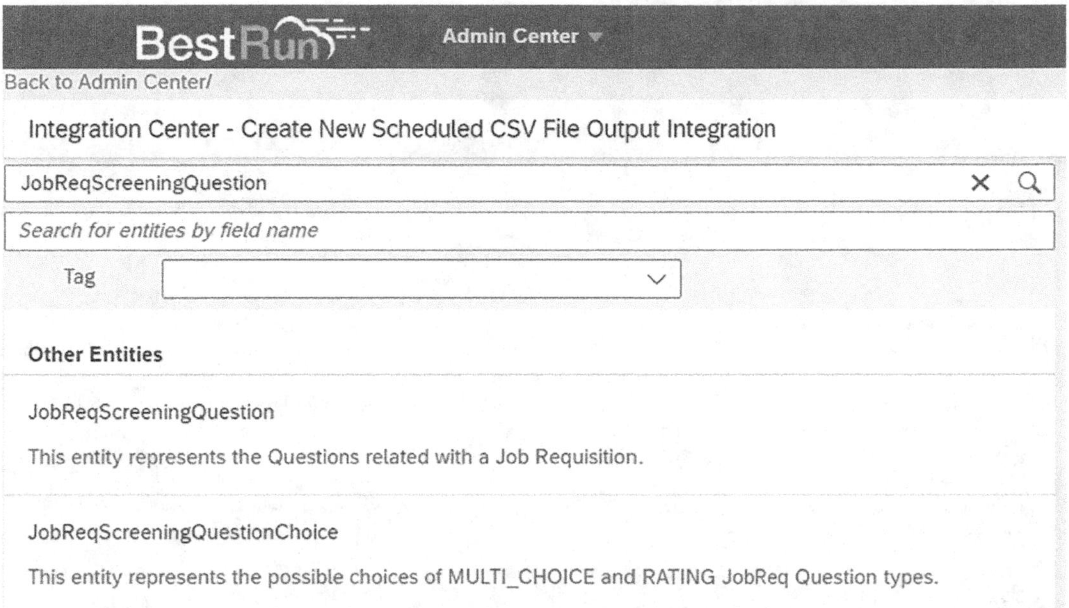

Figure 3-6. *Starting a Query from JobReqScreeningQuestion*

We will then choose the fields which we want to include in the output as well as the entities that we will navigate to. In Figure 3-7, we navigate to both the question choices and the requisition. Had we chosen JobRequisition as the starting entity, then we would choose JobReqScreeningQuestion as the entity to navigate to. The Navigations section always shows the entities that can be immediately navigated to from the starting entities.

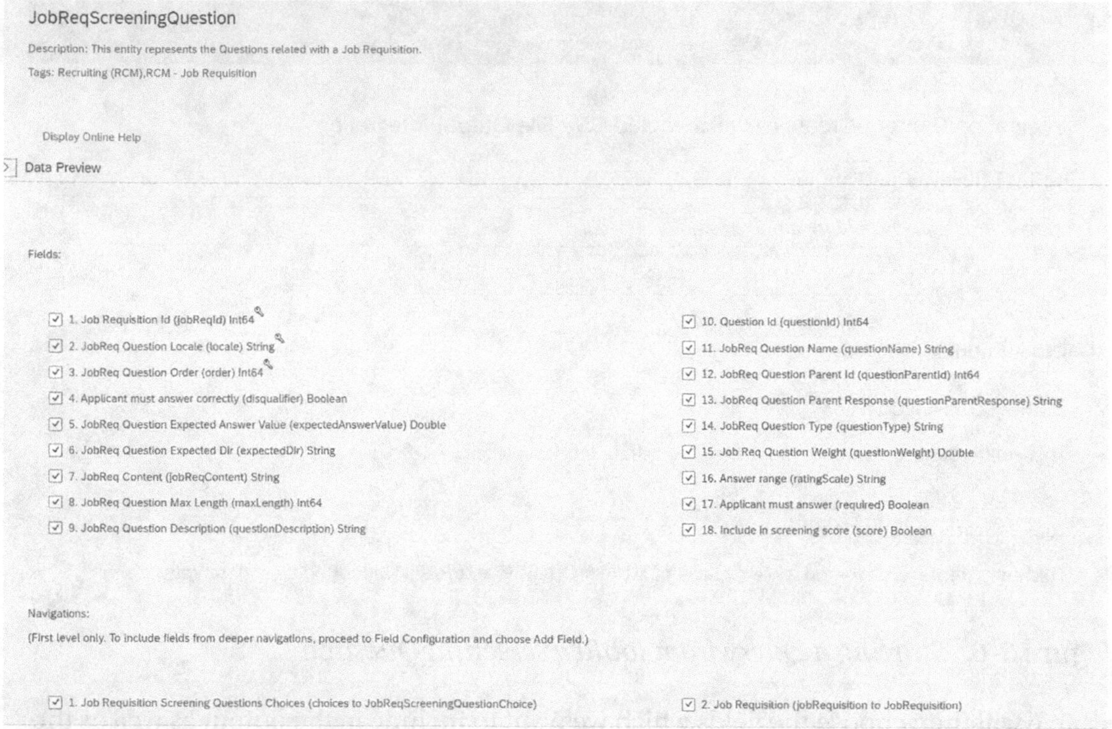

JobReqScreeningQuestion

Description: This entity represents the Questions related with a Job Requisition.

Tags: Recruiting (RCM),RCM - Job Requisition

Display Online Help

> Data Preview

Fields:

- ☑ 1. Job Requisition Id (jobReqId) Int64
- ☑ 2. JobReq Question Locale (locale) String
- ☑ 3. JobReq Question Order (order) Int64
- ☑ 4. Applicant must answer correctly (disqualifier) Boolean
- ☑ 5. JobReq Question Expected Answer Value (expectedAnswerValue) Double
- ☑ 6. JobReq Question Expected Dir (expectedDir) String
- ☑ 7. JobReq Content (jobReqContent) String
- ☑ 8. JobReq Question Max Length (maxLength) Int64
- ☑ 9. JobReq Question Description (questionDescription) String

- ☑ 10. Question Id (questionId) Int64
- ☑ 11. JobReq Question Name (questionName) String
- ☑ 12. JobReq Question Parent Id (questionParentId) Int64
- ☑ 13. JobReq Question Parent Response (questionParentResponse) String
- ☑ 14. JobReq Question Type (questionType) String
- ☑ 15. Job Req Question Weight (questionWeight) Double
- ☑ 16. Answer range (ratingScale) String
- ☑ 17. Applicant must answer (required) Boolean
- ☑ 18. Include in screening score (score) Boolean

Navigations:

(First level only. To include fields from deeper navigations, proceed to Field Configuration and choose Add Field.)

- ☑ 1. Job Requisition Screening Questions Choices (choices to JobReqScreeningQuestionChoice)
- ☑ 2. Job Requisition (jobRequisition to JobRequisition)

Figure 3-7. *Choosing Fields and Navigations from Starting Entity*

Choose a name for your integration as in Figure 3-8 and click Next. As we're only testing the field output, we don't need to worry about the file type or other information.

Figure 3-8. *Choosing an Integration Name and Output Options*

The Configure Fields screen in Figure 3-9 shows a preview of the data output. We will show a fraction of the functionality here as we will be doing a simple column output.

Figure 3-9. *Data Output Preview*

In order to add JobRequisition and JobReqScreeningQuestionChoice fields, we will add columns using the "+ Add" button followed by "Add Field" from the menu. Fields will be chosen from the Entity Tree View option on the left side. Scroll down to JobRequisition to add JobRequisition fields and JobReqScreeningQuestionChoices to show the question choices for multiple choice. Choosing an entity will make the fields or navigation options for that entity appear to the right. Figure 3-10 shows the addition of the Option Label field via the choices navigation. To add multiple fields without leaving this screen, check "Add Another" at the bottom left and click "Add Association" after selecting each field. Click Cancel when you are done.

Figure 3-10. *Adding Fields to Integration Output*

When we look at the output in Figure 3-11, we will see something odd. There is only one row for each of the multiple-choice questions. We would expect multiple, given that there are multiple choices. The Question Id column shows the Question Id that matches the GUID from the question library. For each question, there is only one row.

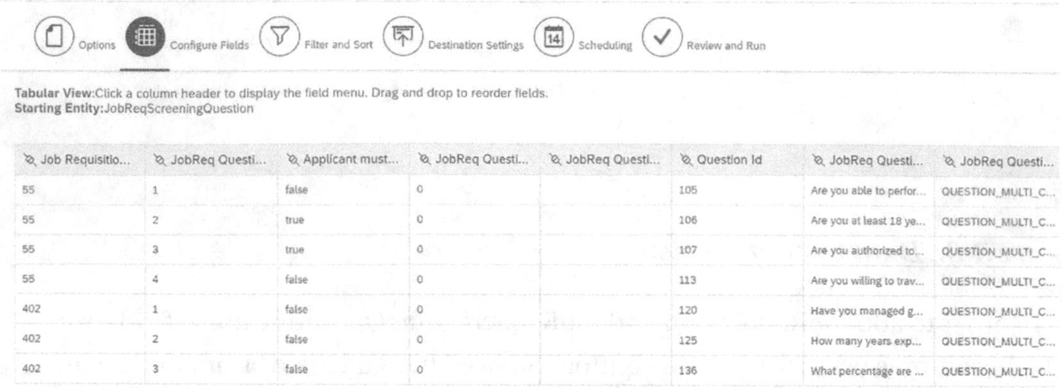
Options
Configure Fields
Filter and Sort
Destination Settings
Scheduling
Review and Run
Tabular View:Click a column header to display the field menu. Drag and drop to reorder fields.
Starting Entity:JobReqScreeningQuestion
Job Requisitio... | JobReq Questi... | Applicant must... | JobReq Questi... | JobReq Questi... | Question Id | JobReq Questi... | JobReq Questi...
55 | 1 | false | 0 | | 105 | Are you able to perfor... | QUESTION_MULTI_C...
55 | 2 | true | 0 | | 106 | Are you at least 18 ye... | QUESTION_MULTI_C...
55 | 3 | true | 0 | | 107 | Are you authorized to... | QUESTION_MULTI_C...
55 | 4 | false | 0 | | 113 | Are you willing to trav... | QUESTION_MULTI_C...
402 | 1 | false | 0 | | 120 | Have you managed g... | QUESTION_MULTI_C...
402 | 2 | false | 0 | | 125 | How many years exp... | QUESTION_MULTI_C...
402 | 3 | false | 0 | | 136 | What percentage are ... | QUESTION_MULTI_C...

Figure 3-11. *Requisition Question Output Example*

Integration Center is designed for simple outputs, and we have demonstrated a limitation. We will use Postman to get a more complete picture of the JobRequisitionScreeningQuestion and JobReqScreeningQuestionChoice entities.

Requesting Job Requisition Question Choices Using Postman

In Chapter 1, we used Postman to issue an ODATA call to SAP SuccessFactors to retrieve some job requisition information. In this example, we're going to pull entities that are associated with that job requisition. We will request questions associated with a job requisition in addition to the question choices for multiple-choice questions.

Log into SAP SuccessFactors Recruiting test or demo environment, and find a job requisition that has more than question associated with it. There should be at least one question that is a multiple-choice question. If you don't have the data, either create a new requisition with questions or add questions to an existing requisition. Take note of the requisition ID.

Open the Postman application and ensure that the environment that you created in Chapter 1 is selected as shown in Figure 3-12. A common mistake with multiple environments is to think that the data or the http request is wrong, when actually the wrong environment is being used.

Figure 3-12. *Selection of Environment in Postman*

In the top left of the screen, click New or File ➤ New Tab to open a new tab and select "Request." Use the request name "GET Job Requisition Questions," select the Job Requisitions collection that you created in Chapter 1, and click "Save to Job Requisitions." Enter the URI of the GET request in addition to the Params, as shown in Figure 3-13.

Figure 3-13. *Parameters and URI to Retrieve Job Requisition Questions*

Don't forget to select the Authorization tab and complete the same user and password as in Chapter 1, using the environment variables {{Username}} and {{Password}}.

The URI for my requisition is *{{Host}}/JobReqScreeningQuestion?$format=json&$select =jobReqId,jobReqContent,questionId,questionName,questionType&$filter=jobReqId eq 55*.

Let's examine the preceding parameters:

$format=json

In Chapter 1, the message body of the ODATA response used the XML format. The JSON format is an open standard lightweight data interchange format that is "self-describing," meaning it is easy for both machines and humans to read. It's very commonly used for web applications to communicate with a server. You will notice in the output that it is much easier to read.

$select=jobReqId,jobReqContent,questionId,questionName,questionType
$filter=jobReqId eq 55

This specifies the job req ID as being 55. For your SAP SuccessFactors system, it will likely be different. In Chapter 1, we used the URI host/JobRequisition?internalStatus= Approved. So why do we need to use the filter parameter now? It's because we are not accessing the Job Requisition directly but via the JobReqScreeningQuestion.

Run the query by clicking Send and let's look at the results in the JSON format. It should look like Figure 3-14.

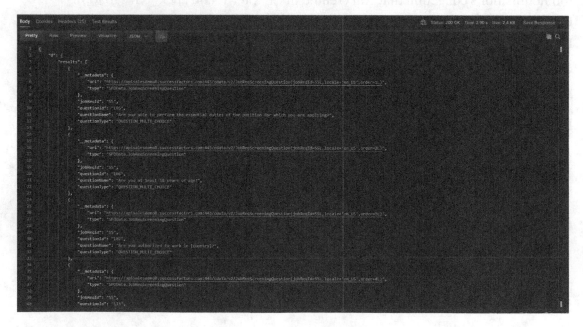

Figure 3-14. *ODATA Get Response in JSON Format*

For each result, there is a metadata section indicating the URI to retrieve the JobReqScreeningQuestion entity as well as the fields that we specified for inclusion using the *$select* parameter.

Now that we have retrieved JobReqScreeningQuestion entities, let's choose one and retrieve the possible question responses. Pick a result that has a questionType of QUESTION_MULTI_CHOICE and note the questionId. Let's now build a query to retrieve the question choices.

As you did previously, create a new Request, name it GET Job Requisition Question Choices, and choose the Job Requisitions collection before saving.

We have included a small number of JobReqScreeningQuestion fields in the results. We are now going to change the parameters to include all fields in addition to the JobReqScreeningQuestionChoice entity.

Change the Postman parameters to the following:

- Uncheck "$select" as a parameter.

- Add the $expand parameter with value "choices."

The URL should now look something like *{{Host}}/JobReqScreeningQuestion?$expand= choices&$filter=jobReqId eq 55&$format=json*.

When you click "Send" to execute the GET request, you should see results like Figure 3-15, depending on the data in your system.

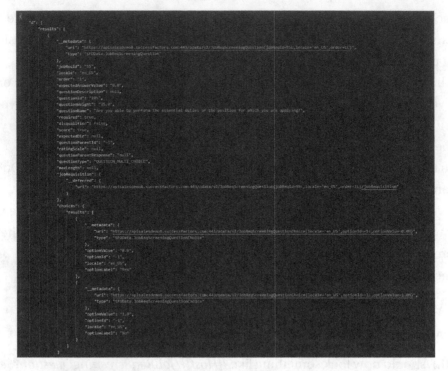

Figure 3-15. *JSON Results of JobReqScreeningQuestion, Including Choices*

Each question is represented by the following within the results tag:

- **URI** – The URI to directly access the entity

- **Type** – The type of ODATA entity

- A list of entity fields and values

- A URI for the JobRequisition

- A nested list of choices for each question, including the URI of the choice, type JobReqScreeningQuestionChoice, and a list of fields and values for each choice

You will notice that the entity referenced in the choice URI is JobReqScreeningQuestionChoice, yet we used the $expand value of "choices" and the results are also included for choices. This is because although JobReqScreeningQuestionChoice is the *entity*, the *navigation* value from JobReqScreeningQuestion to JobReqScreeningQuestionChoice is "choices."

In our first example using the URI *{{Host}}/JobReqScreeningQuestion?$format=json& $select=jobReqId,jobReqContent,questionId,questionName,questionType&$filter=jobReq*

Id eq 55, we included a small number of fields in the results. The second URI of *{{Host}}/ JobReqScreeningQuestion?$expand=choices&$filter=jobReqId eq 55&$format=json* included all fields for the question, in addition to all fields for each choice. Why wouldn't we just use the second URI, given that it's simpler and produces all the information we need?

In a production system, large amounts of data will be returned, reducing the response time. By using $select to return just enough information to apply business logic, the returned URI can then be used to pull data for the desired questions. For example, if we only need question choice information for questions with the words "electronics" or "engineering" in the question name, we would need to retrieve the questionName. The reason is that questionName is not a filterable field. Any filtering needs to be handled by the middleware.

The $filter parameter has some very useful operators such as eq for equal to, ne for not equal to, gt for greater than, and lt for less than, in addition to And, Or, Not, etc.

ODATA V2 also has some useful filter functions such as substringof(string to match, field to match against). An example of this is */JobRequisitions$filter_ substringof('En gineer', jobReqLocale/jobTitle) eq true*. This only includes requisitions with the string 'Engineer' in the job title.

For a full list of operators and function, see `www.odata.org/documentation/odata-version-2-0/uri-conventions/`.

Adding a Question to a Requisition Using Custom Configuration

As we said in the introduction, this book will not teach you how to build middleware integration with tools such as SAP Cloud Platform Integration (CPI). Though integration developers will benefit from this book, the intent is to teach clients and functional consultants how to build proof of concepts and correctly describe the required specifications of an interface. Here, you will see the following:

- The process by which a question is added to a job requisition

- An example of a custom metadata framework(MDF) object to configure the creation of questions on requisitions

- An example of a Postman call to add a question with multiple choices to a requisition

Question to Requisition Addition Process

- Trigger the addition of questions to a requisition.

- Use MDF object to configure which questions to add to which requisitions.

- Add questions to requisitions.

- Update requisition to indicate the update was successful.

Trigger the Addition of Questions to a Requisition

Any questions to be added to a requisition must be done before the requisition is posted. We don't want people applying to the job before the questions are added. The automated adding of questions should also be done before a recruiter has added any questions. Otherwise, they may do unnecessary effort.

There are two main methods of triggering the addition of questions to requisitions:

- Using the Update of Job Requisition event in the Intelligent Services Center (ISC).

  - ISC enables events predefined by SAP SuccessFactors to trigger actions such as an outbound Integration Center call.

  - Since the event will be triggered every time a requisition is created or updated, the filter in Integration Center can be used to determine which events should trigger an addition of questions.

  - The use of ISC will be covered in more detail in Chapter 12.

- Use of call from Cloud Platform Integration (CPI) that is scheduled on a regular basis. This will check for requisitions that haven't had the questions automatically added.

Between the two approaches, we prefer the real-time trigger from ISC. It adds the questions at the earliest time when the requisition is created. It also reduces CPI overhead by not making the necessary frequent calls to SAP SuccessFactors.

To simplify the integration and reduce the overhead of reading JobReqScreeningQuestions, we recommend adding a simple Boolean field to the requisition that indicates that the questions have been added. This can then be used as a filter criterion when determining whether questions need to be added. As this is only used for integration, this should not be visible to recruiting users.

Custom MDF Object to Determine Questions to Add to Requisition

The main reason for having logic to always add certain questions to certain types of requisitions is to automatically disqualify applicants and to ensure those disqualifying questions are consistently applied. Application fields don't automatically disqualify applicants (though they could, using the power of integration.) The other downside of application fields is that the differentiation of applicant fields leads to additional application templates and requisition templates. Application fields can be differentiated by hiring country and whether applicants are externals or existing employees. Downsides of using requisition-specific questions are

- They're not reportable (though this can be done from Integration Center).

- They can't be used as email/offer letter tokens.

- They can't later be used in business rules.

- They can't be used as columns when viewing a list of application.

- They are always presented to internal applicants as well as externals.

Given that disqualifying questions should be consistent, should not adversely affect certain groups of people, and should be really necessary for the job, what kind of criteria should be used to determine questions? The following are some examples:

- **Requisition template** – For example, professional vs. internships.

- **Country** - Questions that determine country-specific hiring eligibility or to rule out questions that can't be asked.

- **Job family or role** – These are usually less hard and fast disqualifying questions. Example would be necessary driving licenses or certifications.

Figure 3-16 shows an example of a custom MDF object that is used to automatically add questions based on the requisition template. If you have very few questions to default and the logic is simple, why not tell the developers which questions to default rather than adding a custom MDF object? Requirements change over time with processes. A system admin needs to be able to update rules rather than asking developers.

69

RCM Requisition Question Defaulting:

* externalCode	Global_18Years
externalName	18 Years all reqs
* effectiveStartDate	05/24/2021
Requisition Template ID	291
Question Order	001
Question ID (GUID)	893
Disqualifying Question	Yes
Required Question	Yes

Figure 3-16. *Custom MDF Object to Default Requisition Questions*

In the preceding example, we've used the externalCode of "Global_18Years?" to make it easier to identify this question and that it's valid for all requisition templates. Note that we've left Requisition Template ID blank to indicate that this question is valid for all requisition templates. We're not specifying the question type or possible answers here. Instead, we are referencing the Question ID, which is the GUID for the question, which can be found by exporting the question library.

Maintaining the previous configuration will be an ongoing process performed by recruiting administrators. For this reason, it has to be secured using role-based permissions (RBP) per Figure 3-17.

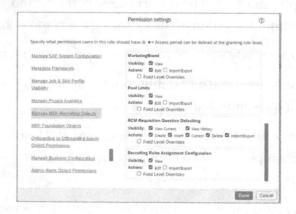

Figure 3-17. *Example Role-Based Permission for Question Defaulting Configuration*

Figure 3-18 shows the high-level steps to default questions on a job requisition.

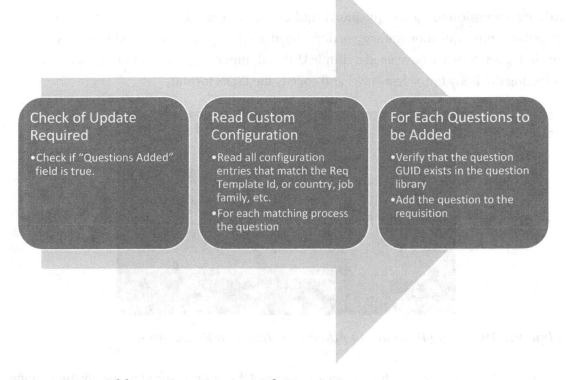

Figure 3-18. *Adding a Question to a Job Requisition*

The following needs to occur when the integration is triggered:

- Check if the requisition questions have already been automatically added.

- Read the custom configuration to determine which questions to add and whether they are disqualifying questions and if they are required.

- For each question that is relevant for the job requisition

  - Read the question and choices (if multiple choice) from the question library.

  - Create the job requisition screening question against the requisition.

  - Create the possible answers (if multiple choice) against the newly created job requisition screening question.

Example Postman Call to Add Job Requisition Screening Question

Adding a question to a job requisition is quite straightforward. Since we are using a question from a question library, we don't need to specify the question choices. As we can see in Figure 3-19, we have used a simple URI endpoint of /upsert?format=JSON. After selecting the Body tab, select "raw" and specify the JSON format.

Figure 3-19. *Using Postman to Add a Question to a Requisition*

The URI and type within the metadata section of the body are used to specify the specific entity and type. The URI can include a parameter indicating an existing entity for update. In this case, it's unnecessary as we are adding a new question using the UPSERT method.

In Figure 3-20, we can see the results. A JobReqScreeningQuestion has successfully been added to requisition 1365.

Figure 3-20. *Results After Adding Job Screening Question*

If we were to add multiple-choice questions that weren't from an existing library, this would necessitate the addition of choices as the entity JobReqScreeningQuestionChoice. This is not a recommended approach.

Conclusion

You have now learned how to choose the right tool for the automation of recruiting. This will serve you well in understanding client needs and how they can be solved for. This will be invaluable in deciding whether a requirement can be solved using business rules, the Integration Center, or through third-party middleware such as CPI.

You've also learned when to use job requisition questions vs. application fields in capturing applicant information. To support the defaulting of job requisition questions, you learned how the entities of JobRequisition, JobReqScreeningQuestion, JobReqScreeningQuestionChoice, and JobReqQuestions fit together.

In building on what we learned in Chapter 1, we dived deep into our first ODATA Upsert to add questions to job requisitions. For this, we used the Postman tool. In doing so, we learned how to formulate GET requests using the JSON format.

You also learned how to use a custom MDF object to configure the addition of job requisition questions.

In the next chapter, we'll talk about automating postings and will also revisit ODATA by showing how to display posting data in WordPress.

CHAPTER 4

Candidate Attraction

In the previous chapter, we looked at how automation tools can be used to reduce the time that recruiters spend on admin tasks like adding the same application questions to a requisition again and again. One of the goals of recruiting technology is to free up recruiter time to better engage with future employees. Chapter 5 will discuss engagement of candidates on the career site, how to automatically interact with site visitors and applicants once they have taken the first step and visited the site and given some information about themselves. What about bringing visitors to the site in the first place?

In this chapter, we will look at how to bring visitors to your career site, either directly to postings or to specific content.

Overview of Candidate Attraction

Most organizations need to work hard to find and hire the best talent. Very few employers have the pleasure of being considered the employer of choice in their industry or part of the world. Even those lucky few organizations still have specialist jobs where the right marketing method and medium is needed to attract that type of talent to fill those specialist roles.

Talent marketing is the strategy, process, and methodology of getting the right people to want to work at your organization. This is similar to the process of attracting and retaining customers. It includes but is not limited to:

- Attracting candidates with diverse backgrounds and experiences

- Being present and having a voice where communities of talent gather: web forums, fairs, colleges, websites, alumni associations, etc.

- Making persuasive job postings that resonate with candidates

© Anand 'Andy' Athanur, Mark Ingram and Michael A. Wellens 2022
A. A. Athanur et al., *Innovative SAP SuccessFactors Recruiting*, https://doi.org/10.1007/978-1-4842-7425-5_4

- Posting jobs where you can reach candidates with the right experience

- Leveraging the power of employees as employer ambassadors, networkers, and candidate referrers

- Reinforcing the Employer Value Proposition (EVP, or "Why work here?") in everything you do and say

- Ensuring that sources, recruiters, hiring managers, and employees can articulate the EVP with one voice

- Creating a pleasing-to-use career site with an engaging experience that lasts beyond the initial application

- Providing value to potential candidates such as job matching, career tips, and information from employees on what it's like to do their job

Talent marketing is a huge task and needs a strategy that is tailored to the right talent and competitive hiring landscape. Technology can automate the process and amplify your message to a greater audience. If your Employer Value Proposition isn't authentic and compelling, no amount of technology will fix it.

Strategies and Tools for Increasing Candidate Attraction

Figure 4-1 shows a subset of places that can bring site visitors to look at job postings. Starting with the job postings in green at the center and working outward, you can see the different site types of channels in yellow where the postings can be seen to the specific examples of those channels at the outer edge of the diagram in white. There are an overwhelming number of options, and the combination of those chosen will be based on where your targeted talent visits, your hiring volume for a segment of talent, and of course budget. The intent of this book is to support technical innovation and automation within recruiting rather than teach talent marketing.

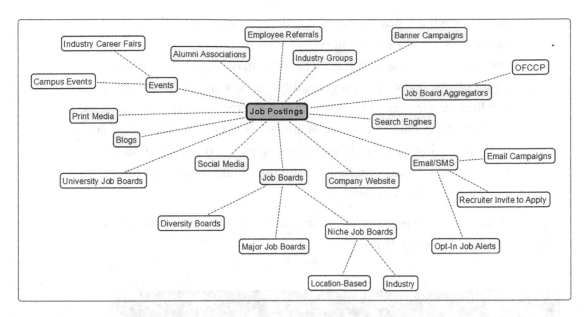

Figure 4-1. *Example Sources of Candidates*

Talent marketing is distinctly separate from sourcing, where sources reach out to passive candidates (read, not looking for a job) using Google resume searches, LinkedIn, social media, etc. The pool of candidates within your Applicant Tracking System (ATS) is a source of candidates, though you won't typically find passive candidates there. Recruiting is the process of managing the recruiting activities on applicants for a particular requisition. The role of sourcing and recruiting may be blended, especially in smaller organizations.

The initial drive of candidates to postings is just one part of the candidate life cycle, as shown in Figure 4-2. Shown left to right, you can see the candidate life cycle:

- Attract, where job postings are published to different channels.

- Personalize, where candidates can opt in to job alerts and marketing to receive notification of new job postings relevant to their interests.

- Engage, where the candidate interacts with the career site, receiving information through various channels such as chatbots, SMS, email, etc.

- Convert is the stage where a candidate becomes an applicant by proactively specifying interest in a specific position.

- Recruit is what is traditionally considered applicant tracking. Applicants are screened, interviewed, hired, or dispositioned. Previous sourcing efforts pay off because existing pools of qualified talent can be invited to apply. In addition, AI searching and matching can find existing candidates that match the requisition requirements. Recruit also becomes a source of future talent engagement because candidates that were qualified but edged out by other candidates are considered "silver medalists" and put in talent pools for future sourcing and engagement.

- Develop is both the support of internal mobility and the use of employees to supply referrals.

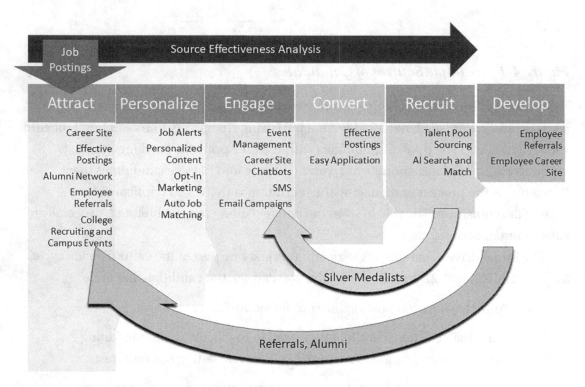

Figure 4-2. *The Candidate Life Cycle*

Beyond talent marketing through the distribution of job postings, the candidate life cycle includes candidate relationship management (CRM), which is the process of automatically engaging with candidates using self-identification of interests, talent pool management, and email campaigns as described in the Personalize and Engage phases earlier.

No recruiting solution can solve all problems. SAP SuccessFactors Recruiting sits within an ecosystem of complementary solutions that integrate with it. You can find many third-party solutions on the SAP Store (see Figure 4-3) at `https://store.sap.com/dcp/en/search/recruiting`. Note that solutions don't have to be on the SAP Store to work with SAP SuccessFactors Recruiting.

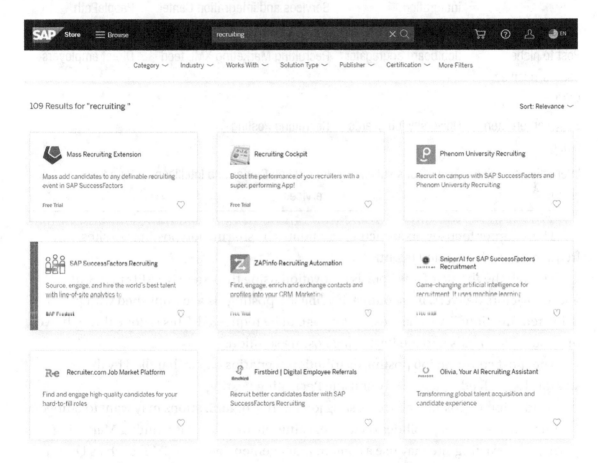

Figure 4-3. *The SAP Store*

Relevant Business Scenarios for Candidate Attraction That Require Integration

Table 4-1 shows a list of common talent marketing and sourcing needs, third-party solutions that are available to meet that need, the type of integration required with SAP SuccessFactors Recruiting, and example vendors. Note that the listed vendors are just examples and not recommendations.

Table 4-1. *Examples of Candidate Attraction Business Scenarios Requiring Integration*

Need	Solution	Integration	Example Vendor
Sourcing alumni	Alumni solution integration	Candidate import using Intelligent Services and Integration Center, postings via XML feed	Enterprise alumni, PeoplePath
Post to niche boards, such as OFCCP	Job board aggregator	Recruiting Marketing XML feed	Direct employers
Early career/intern hiring	University job boards	Recruiting Posting	
Interactive sourcing via SMS	AI chatbot solution	Integration Center and Intelligent Services	Paradox

This chapter focuses on attracting candidates by sharing job postings to sites frequented by the needed talent.

Though the theme of this book is innovation using the extensive APIs that SAP SuccessFactors delivers, most outreach using job postings is accomplished with pre-delivered functionality by SuccessFactors or can be requested. This is done through two methods: XML feeds and the Recruiting Posting solution.

The most common job posting distribution scenarios can be handled by the standard XML feed functionality or using Recruiting Posting.

In addition to posting jobs to existing job boards, organizations may want to embed postings and search capabilities on their existing site outside of Recruiting Marketing (RMK). The existing site may use a content management system (CMS) such as Drupal or WordPress. At first glance, it may seem logical to work with the organization's web development team to make ODATA calls, pulling requisition, and posting data. However, XML feeds are a much better choice. The reasons are

- It takes the work out of building an ODATA REST call and parsing the XML or JSON results because XML feeds are designed as simple posting information with just the right information.

- No API user is needed to make the call. It's a publicly available XML feed. This reduces complexity and decreases the number of third parties making calls to SAP SuccessFactors.

- The feed can be prefiltered using rules if the postings need to be categorized by specific countries, divisions, job functions, etc.

- The apply URL can have a campaign code parameter assigned specific to the rule (the UTM_Campaign parameter), to give more accurate advanced analytics information.

Case Study: Reposting Using WordPress

An enterprise software consulting company has a blog on its website that covers many topics, including SAP SuccessFactors Recruiting. WordPress is the platform used. When blog posts are focused on SAP SuccessFactors Recruiting consulting topics, the company wants to embed related job postings within the blog post. The list of postings needs to be updated automatically without editing the blog post. Our example site is shown in Figure 4-4.

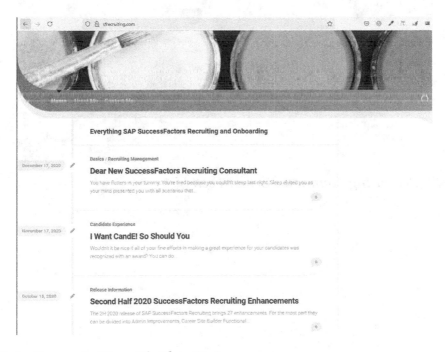

Figure 4-4. *Example Site with Blog Posts*

The following are some key requirements:

- Blog posts aimed at consultants should automatically include a list of relevant postings at the bottom.

- Each job posting should be a linked blog post with a link back to the posting on the main Recruiting Marketing career site. Our talent marketing specialist told us that the links are good for search engine optimization (SEO).

- Our home page normally shows a list of blog posts. We don't want job postings included here as this makes it look like a job board instead of a blog site with useful content.

Figures 4-5 and 4-6 show the Recruiting Command Center.

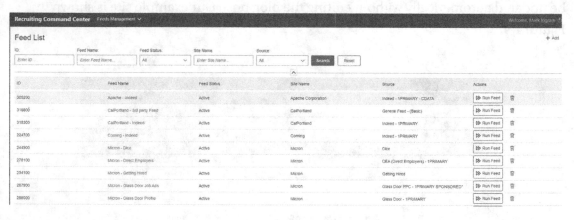

Figure 4-5. *Recruiting Command Center Feed List*

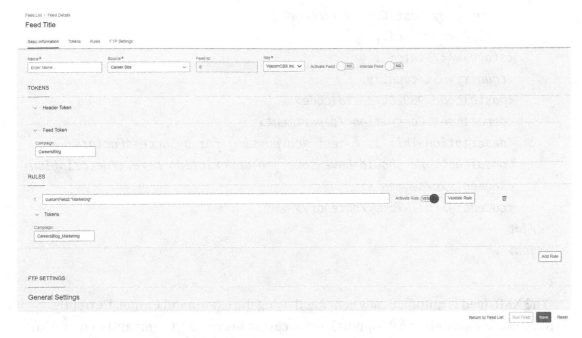

Figure 4-6. *Recruiting Command Center Feed Details*

Example XML for the job posting feed is shown as follows. Note that your XML feed will likely differ.

We don't want all job postings to go to the WordPress site. We want to connect with blog readers to specific job postings related to the career that they care about. In the earlier example, we are using a rule to only include posting titles that include "SuccessFactors Consultant."

```
<?xml version="1.0" encoding="UTF-8" ?>
<source>
<publisher>.myTestRMKSite</publisher>
<publisherurl>jobs.myTestRMKSite.com</publisherurl>
<jobs>
    <job>
        <title>Consultant</title>
        <date>Fri, 25 06 2021 00:00:00 GMT</date>
        <referencenumber>540-en_US</referencenumber>
        <url>https:// jobs.myTestRMKSite.com/job/Consultant-
WA-98020/86338550/?feedId=18876&utm_source=WordPress&utm_
campaign=SomeTestBlog</url>
```

```
    <company>My Test Company</company>
    <city>Edmonds</city>
    <state>WA</state>
    <country>US</country>
    <postalcode>98020</postalcode>
    <department>Consulting</department>
    <description>This is a test job posting for a SuccessFactors
    Consultant. We should have some job description here.</description>
    <segments></segments>
    <category>Full-Time</category>
</job>
<job>
```

The XML feed is automatically generated using the Command Center. Currently certified consultants and SAP support have access to Recruiting Command Center. Most of the XML is self-explanatory. Note some key values:

- **Title** – Posting title

- **Date** – Posting date

- **Reference number** – This is the job requisition ID plus the language locale. There may be multiple postings in different languages.

- **URL** – This is a link for a candidate to apply. It takes the candidate to the posting within RMK. Within the link are two URL parameters that are used by advanced analytics:

  - *Utm_source* – We will use the predefined source of "WordPress."

  - *Utm_campaign* – We will use this to differentiate between blogs. I'm using a test one *SomeTestBlog*.

Once the feed is generated, we can use the feed URL to populate WordPress. On our WordPress site, we are going to do the following:

- Automatically import external job postings, creating a WordPress post for each one.

- List the current job postings dynamically on any blog post that uses relevant shortcodes. Shortcodes are a way of embedding dynamic content within posts.

- Automatically exclude job postings from displaying on the WordPress home page.

Note   This is an example of how to include job postings within your content management system (CMS). Other examples are:

- Create a WordPress plugin to show hot jobs.

- Import job posting feed XML to a non-RMK search page hosted by the customer.

- Explore other prebuilt tools such as the WordPress "WP Job Manager" plugin (`https://wpjobmanager.com`). This can be used to support search pages, set up job alerts, etc. This plugin should be of limited use as you should generally use the native RMK search.

Remember to not reinvent the wheel if SAP SuccessFactors already supports functionality. In every posting technology decision you make, consider whether accurate source reporting via advanced analytics will be supported. The technology used will be mostly driven by the client's in-house tools of choice, such as Drupal or WordPress, homegrown solutions.

This book does not introduce WordPress administration. For this example, we are using an already hosted WordPress site and already existing plugins to avoid complexity. You can also install WordPress locally on your computer to test this scenario.

The following are the WordPress plugins that we are using:

- Ultimate Category Excluder (UCE) `https://wordpress.org/plugins/ultimate-category-excluder/`

- WP All Import Pro `www.wpallimport.com/`

- WP All Import Scheduling Subscription (only needed in a production environment)

Figures 4-7 and 4-8 show the WordPress Admin plugins page. This is where plugins are added and activated. It will initially show the existing plugins that have been loaded.

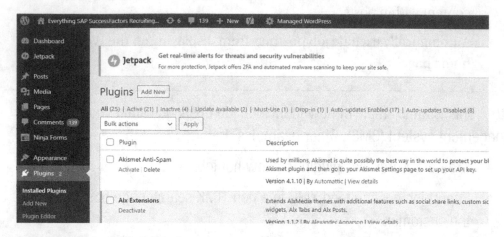

Figure 4-7. *Plugins Within WordPress Admin*

Figure 4-8 shows the results when we search for the Ultimate Category Excluder. The Ultimate Category Excluder (UCE) makes it possible to easily exclude categories from your front page, archives, feeds, and searches. You will need to install and activate this as well as the WP All Import plugin.

Figure 4-8. *Searching for the Ultimate Category Excluder Plugin*

After the plugins Ultimate Category Excluder and WP All Import Pro (or non-pro version, which is free) have been installed and activated, let's put them to work.

First, we want to set up and import job for the job posting XML feed that we previously created in Command Center.

On the main menu, click *All Import* as shown in Figure 4-9. This is the plugin that we previously activated. Then click "*New Import.*" We do this in Settings. When we click Settings on the left side, any activated plugins that have settings will be shown below the Settings menu.

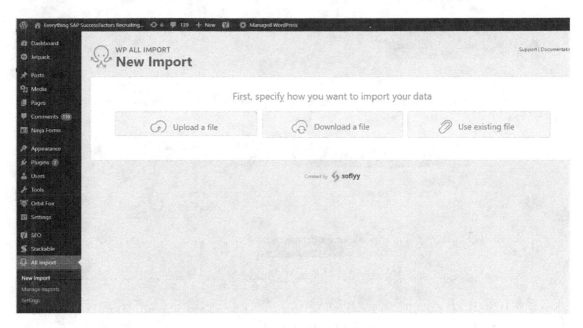

Figure 4-9. *Starting the Initial Setup and Posting Load for WP All Import*

Click "Download a file" and then "From URL."

We will then provide the URL of the job posting XML feed, per Figure 4-10.

Figure 4-10. *Specifying the RMK Posting XML Feed for WP All Import*

In Figure 4-11, we choose to make WordPress posts for each uploaded job entity in the XML. This could also be used to create pages.

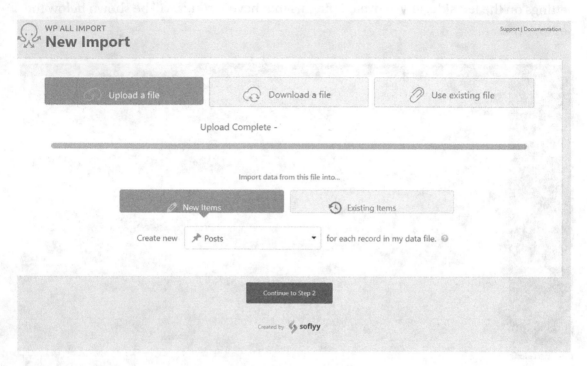

Figure 4-11. *Post Import Options for WP All Import*

Figure 4-12 shows the selection of the data that will be used for a WordPress post. WP All Import will read the XML and show it on the right side, along with equivalent buttons on the left side. Click the XML element that should be used to create a post. In this case, it will be the job as we want a post creating for each job.

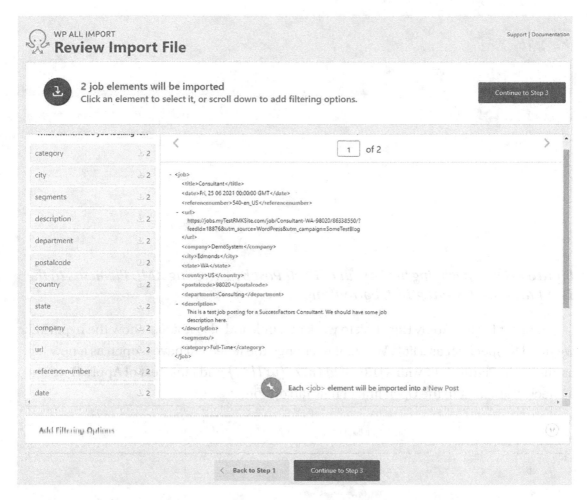

Figure 4-12. *Selecting the Job Element so Each Job Creates a Post*

Figure 4-13 shows how to define a template for a WordPress post, using data from the XML. Specify where the title comes from, the job description, locations, etc. Elements from the XML are selected by choosing them on the right side. Text such as "Location:" can be added to the template and it can also be formatted. When creating a hyperlink, click the chain icon to edit the link properties.

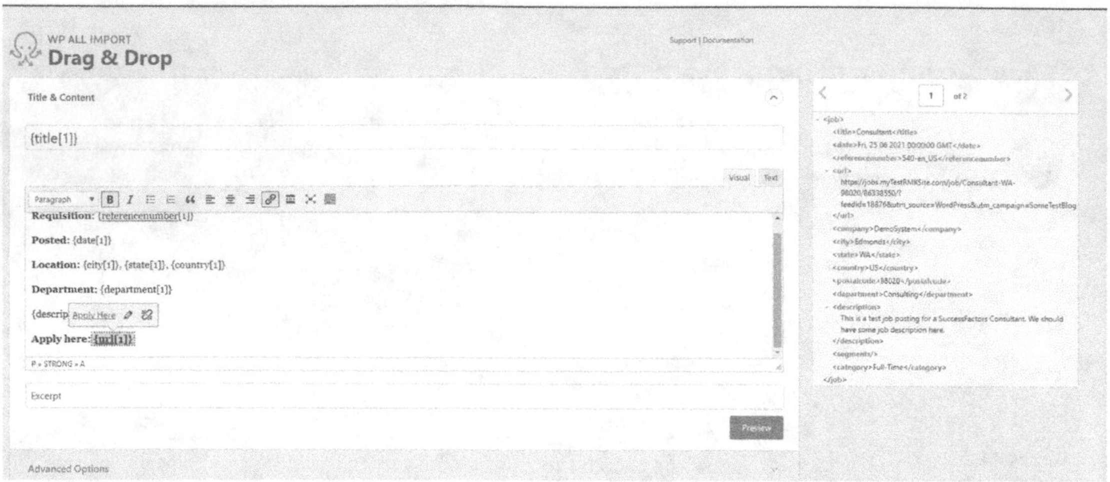

Figure 4-13. *Specifying the Layout of Each Post by Dragging XML Elements to the Text Editor and Adding Text, Formatting*

In order for the Apply Here link to work as a link and also not just show the hyperlink, we need to specify it as a link. We want to change the label and have it open as a new tab, as shown in Figure 4-14 with a URL of $http://\{url[1]\}$ and Link Text of Apply Here and a check mark next to the Open link in new tab option.

Figure 4-14. *Adding the Hyperlink for Applying for a Job*

Click "Preview" to see what the posting will look like, as shown in Figure 4-15. If we were embedding real postings as WordPress posts, we would format the post to make it match the style of the rest of the site.

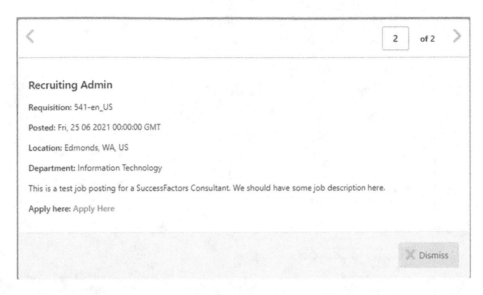

Figure 4-15. *Preview of Automatically Generated Post with Link to Apply*

We then specify the category that the posts will be assigned to, as shown in Figure 4-16; in this case, we are specifying the category JobPosting. This is important for embedding posting links in WordPress posts and for excluding job postings from the blog home page.

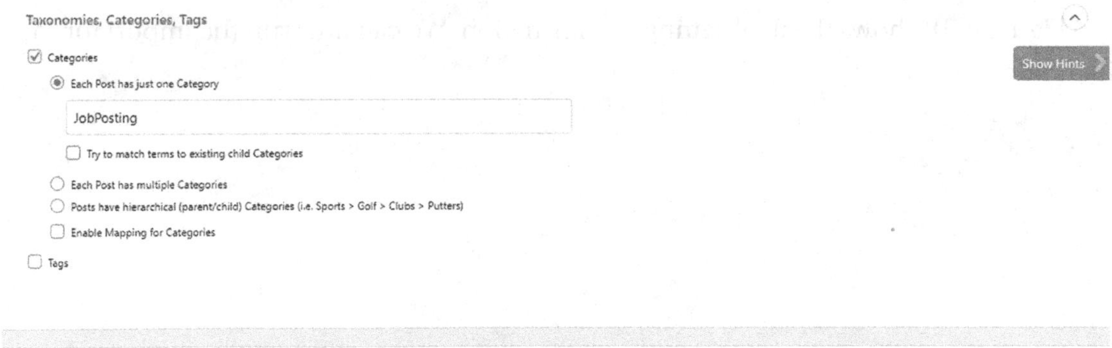

Figure 4-16. *Taxonomies, Categories, Tags*

We then determine which XML element to be the unique identifier of the post and what to do when the job XML is updated or removed in scheduled updates. This can be seen in Figure 4-17.

Figure 4-17. *Import Settings*

Figure 4-18 shows the final settings confirmation. We can now run the import for the first time.

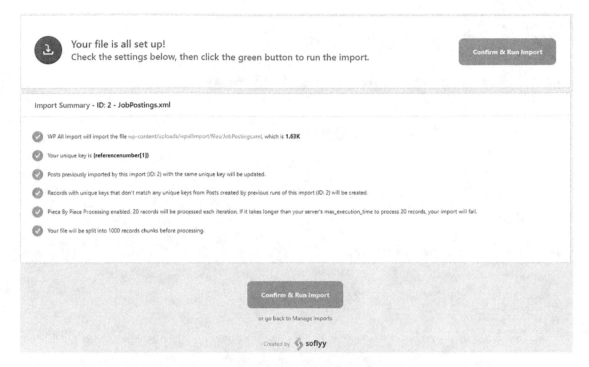

Figure 4-18. *Confirmation of Our Import Settings*

You then get a confirmation message when an import is run successfully. Note that you can see the number of records that have been updated by the successful import, as shown in Figure 4-19.

Figure 4-19. *Confirmation Message*

We then set the import schedule as shown in Figure 4-20. Once a day is recommended, just as we already do with the Recruiting Management to Recruiting Marketing job synch in provisioning.

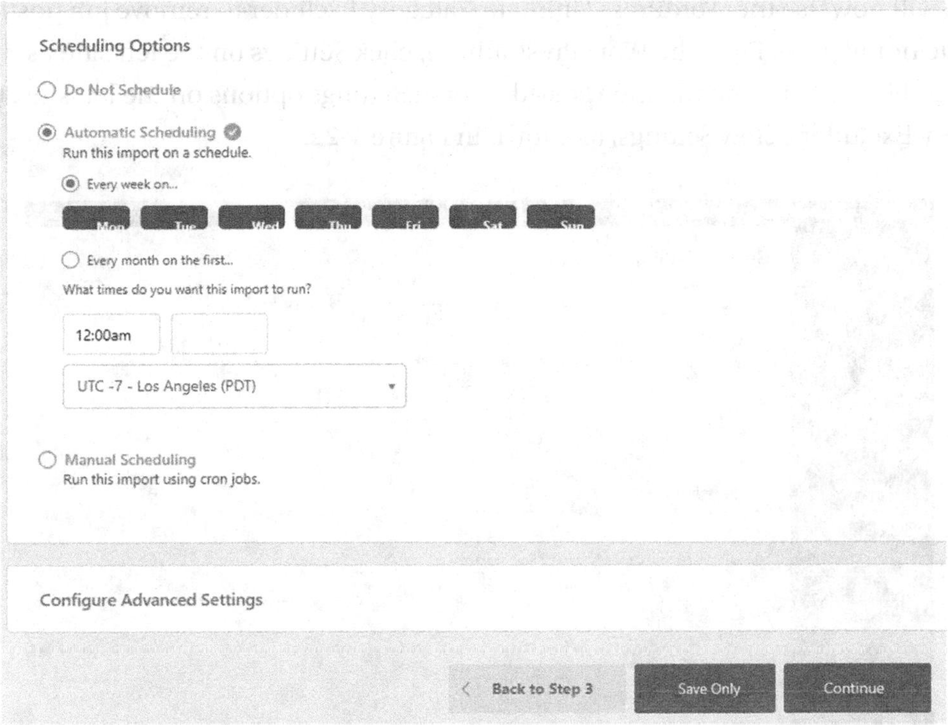

Figure 4-20. *Scheduling Options*

In Figure 4-21, we can see that the two job postings are displayed on the home page as blog posts. We want them to be excluded.

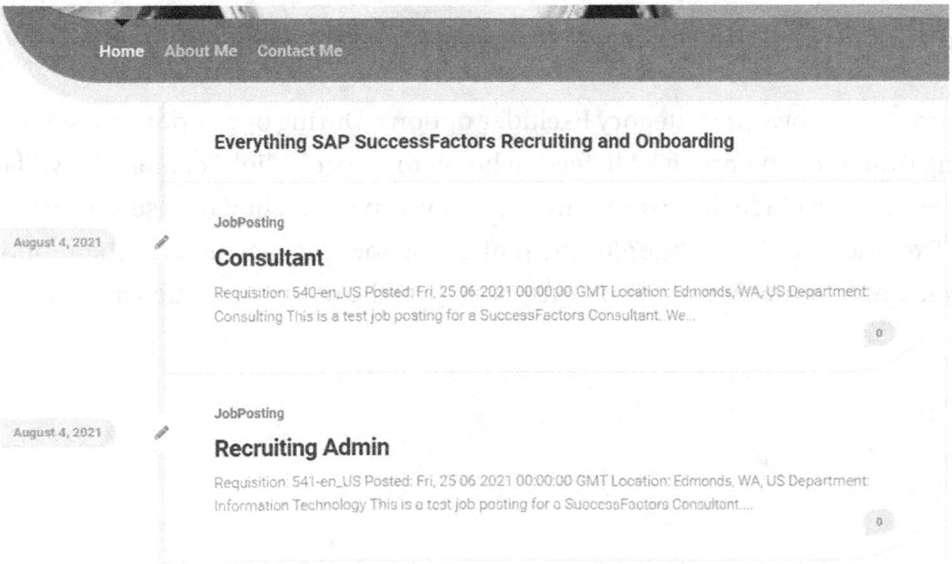

Figure 4-21. *Home Page Showing Job Postings as Blog Posts*

We will now use the WordPress Ultimate Category Excluder to remove job postings from the home page. From the WordPress admin, click Settings on the left side as shown in Figure 4-22. You will then see additional Settings options on the left side. Click Category Excluder below Settings, as shown in Figure 4-22.

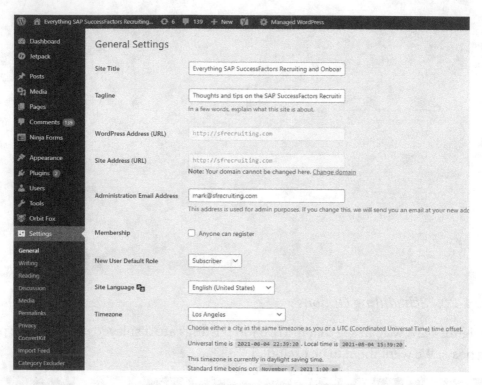

Figure 4-22. *WordPress General Settings and Settings Menu Options*

Figure 4-23 shows the Category Excluder options. During our import, we set anything from the job posting XML feed to be set to category "JobPosting." We will use that category to exclude the postings from the home page as shown by selecting the option Exclude from Front Page? for the JobPosting row and leaving the other options unselected in that row. We will leave the postings in the other areas such as search results.

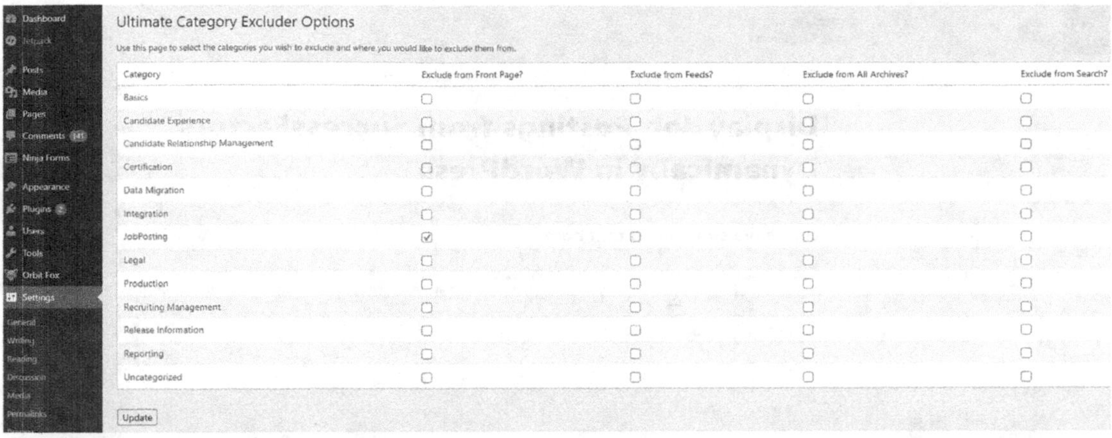

Figure 4-23. *WordPress Category Excluder Options*

After we click "Update" and return to the home page, we will see that the postings disappeared. If they haven't disappeared, you may need to clear your cache. To do so, go to your WordPress admin settings and click "Flush Cache." In this case, Flush Cache was under "Managed WordPress" at the top of the browser page.

The next step is to create a test blog post that dynamically includes a list of job postings. We have used a category called JobPosting to do this. If you want to feed different jobs to different WordPress posts, you can have multiple categories such as "SuccessFactorsRecruitingJobs" and "SuccessFactorsOnboardingJobs." Having multiple categories becomes important where you have blogs that cover multiple talent disciplines, such as programming, marketing, design, engineering, etc., and want to post the right jobs to the right blogs.

We will create a simple WordPress post with a title header, some introductory text, and then dynamically display a list of job postings. This is shown in Figure 4-24.

Figure 4-24. *Simple Example WordPress Post with Shortcode*

Note the use of a *shortcode* at the bottom of the post *[display-posts category="JobPosting"]*. This declares that all WordPress posts with the category "JobPosting" should be listed in the post. We can add other criteria before the closing square bracket, such as *orderby="title".*

After we publish the post, we will see a preview of it on the home page. Click it and we'll see the post as shown in Figure 4-25.

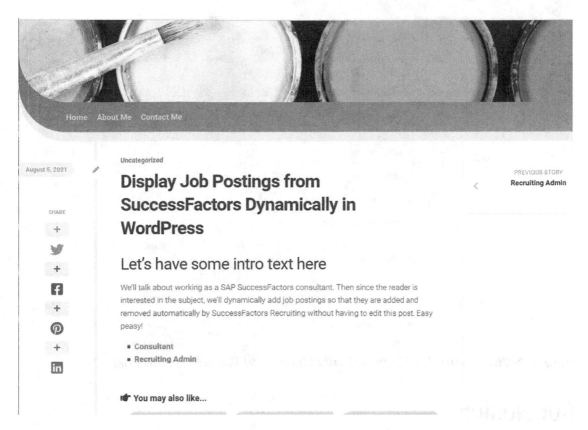

Figure 4-25. *Sample WordPress Blog Post with Dynamically Included Links to Job Postings*

If we click the job title, we will see the WordPress post that was generated from the SuccessFactors XML feed. This is shown in Figure 4-26. Note that if you are re-creating the example using the code in this chapter and click "Apply Here" in the test post, you will receive an error because the example code does not post to a real Recruiting Marketing career site.

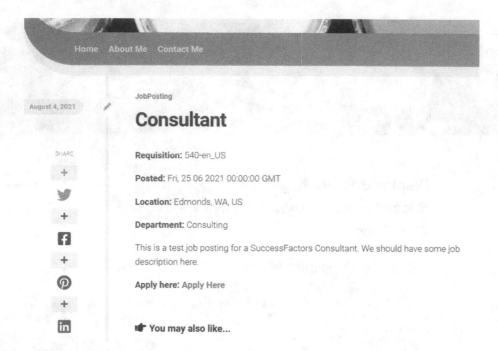

Figure 4-26. *Example of Dynamically Generated WordPress Job Post*

Conclusion

You have now learned the following:

- There are a lot of options regarding talent marketing, and they should be chosen based on the types of talent needed and the source effectiveness.

- Attracting candidates goes beyond job postings (though they're pretty important), but be careful about reinventing the wheel.

- Recruiting Marketing XML feeds are a great solution to be used beyond job board postings. No ODATA calls required!

- WordPress has a lot of plugins that enable low cost and zero code of pulling relevant job postings into your WordPress site.

- Between WordPress, Drupal, and other content management systems, there are a lot of ways of embedding postings in relevant site content.

CHAPTER 5

Enhancing Candidate Engagement

In the last chapter, we learned how to enhance the posting capabilities of SAP SuccessFactors. We now take a look at the next step in the process, candidate engagement. Candidate engagement describes the process of increasing candidate participation in your recruiting activities. In other words, once you have marketed the jobs you have available, how do you help maximize the number of qualified candidates who apply to those postings? If candidates do not see a posting that interests them, how do you keep their interest until the right job posting is available?

Strategies and Tools for Increasing Candidate Engagement

Candidate engagement is critical throughout the recruiting process, and companies have invoked a variety of strategies to increase candidate engagement. Prior to the advent of modern recruiting systems, this was accomplished with recruiters picking up the phone or sending letters or emails to candidates manually. As recruiting systems have advanced, the technology has allowed the amount of manual work the recruiter needs to perform to decrease while increasing the number of candidates a recruiter can engage. For example, out of the box, SAP SuccessFactors provides a suite of candidate relationship management tools designed to help increase engagement. These tools allow recruiters to collect candidate data and target candidates with engaging content. For example, a company at a job fair could create a landing page that collects basic candidate information like name, email, and job interests. After collecting the data, the recruiters can organize data about candidates into talent pools of common interests and/or qualifications. This makes it easier to then target candidates with specific

© Anand 'Andy' Athanur, Mark Ingram and Michael A. Wellens 2022
A. A. Athanur et al., *Innovative SAP SuccessFactors Recruiting*, https://doi.org/10.1007/978-1-4842-7425-5_5

content. Using the email campaigns feature, recruiters can then launch targeted email campaigns to those talent pools. In this manner, recruiters can be more informed about what interests candidates or what might be a good job match and actively engage candidates with relevant content.

The preceding example assumes a candidate is motivated to click a link and fill out information; however, that is not always the case. Sometimes candidates need prompting before they will provide information. This is where creating an enhancement to SAP SuccessFactors such as a chatbot can be useful. Chatbots can interact with a candidate and ask them questions in order to build a candidate profile including contact information and interests. This is performed automatically through programming rather than a recruiter having to manually engage with the candidate using online chat. For example, a chatbot may become active on your recruiting website after a user has navigated on the site for a set period of time and has not applied to a job or created a profile. The chatbot can then prompt the user for name, email, and interests to create the profile on their behalf. In this way, though the candidate did not apply to any specific job, the candidate was still engaged and can be targeted with content later and prompted to apply for positions that could be a match.

Further strategies around candidate engagement focus on keeping candidates engaged after they have already applied or partially applied. For example, if a candidate saved an application but did not submit, a chatbot could send a text to the saved mobile phone on the application reminding the user that the application has not been completed. The chatbot could ask the user if they plan on completing the application or if it should be closed. It can even prompt the user with a questionnaire as to why they decided not to complete the application process. In another scenario, if the application was completed, text messages or emails can be sent periodically at different stages of the application process to let candidates know where they are in the process and what actions need to be taken by the candidate (SAP SuccessFactors can accomplish this using standard functionality).

Even if a candidate has gone through the application process and has not been selected, this does not necessarily mean the relationship is over. Chatbots can also help in this regard. For example, if an automated follow-up email seems too impersonal, a chatbot can be used to perform this task as well. This chatbot could send a message to the user's mobile phone on file seeing if they would be interested in a new posting. It could also touch base and see if they are still interested in the company and encourage them to update their information on the recruiting site.

Figure 5-1 illustrates how the various strategies for engaging candidates can be used in conjunction to work toward the common goal of getting qualified candidates to complete applications. Keeping these business scenarios and strategies in mind, we can now take a look at what data and tools are available to create enhancements and automations that increase candidate engagement.

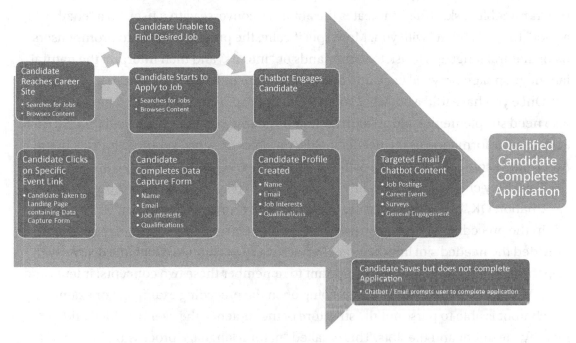

Figure 5-1. *Collaborative Automated Methods for Engaging Candidates*

Case Study: Career Site Chatbot

At this point in the chapter, we should have an understanding of why we would be motivated to create a chatbot and what tools and data are available in SAP SuccessFactors to support creating one. Now let's take an in-depth look at a specific scenario and see how we would go about building one.

In this section, we will expand on the scenario described in the beginning of the chapter where a candidate is unable to find a job and does not apply. After a set period of time, the career site would prompt the candidate for more information about what kind of job they are looking for and collect some basic information about the user. The end goal would be to create a candidate profile in the SAP SuccessFactors system so that recruiters and/or other system automations can engage with the candidate.

Introduction to Chatbots

Before we dive into creating our chatbot automation, let's review some fundamental concepts about chatbots.

The word "chatbot" is the combination of the word "chat" and "bot." A "bot" is a set of automated configuration and/or programming that performs a repetitive or programmable task. "Chat" indicates the ability to converse with a user. So a "chatbot" is a "bot" that can "chat" with you. More specifically, the program is able to prompt users using natural language to ask for commands or "intents" and then interpret the natural language spoken or typed back into one of the commands/intents given to it.

Once you have told a bot what you want to do (e.g., a command or an intent), it may also need supplemental information included with the command. You can think of this additional information as the needed fields/variables or "slots" required for execution. An example some of us may use with our smartphones can be seen in the proceeding:

User: Hey chatbot, add currently playing song to playlist "Mike's awesome playlist."

Chabot: OK, I've added the song to playlist "Mike's awesome playlist."

In the preceding example, the user had an intent of adding a song to a playlist and provided the needed slot information of what song to add and to which playlist. As we step through our case study, it is important to remember these two concepts: intents and slots. There's also a lot of other things going on in the preceding example. For example, the chatbot is able to parse out the structure of the sentence the user provided and identify the intent and the slots. This is called "natural language processing." The good news is there's a multitude of competing cloud chatbot services available out there to take care of that part for you. Since there are so many services available, we can't get into the specifics of how each works. The main thing you will have to understand is regardless of which service you use to build your chatbot, you will need to know what your intents and slots will be. This will allow you to define what specifically your chatbot will do and, in turn, what APIs and data objects the chatbot will need to invoke on SAP SuccessFactors.

Systems Architecture

Figure 5-2 shows us an example systems architecture for a chatbot solution. By now, you've probably realized that unless you want to develop your own natural language processing program, you will need to use a cloud chatbot service to handle processing the input and output of intents and slots with the candidate. Since the candidate is on the career site, this site will serve as our main user interface.

Figure 5-2. *Example Chatbot Solution Architecture*

Note   *If you use SAP SuccessFactors Recruiting candidate experience (also known as Recruiting Marketing), you will notice the Career Site Builder functionality does not come with a chatbot component. You may have to create the chatbot front-end layer using a custom component or create your own site outside of the Career Site Builder in order to create a chatbot.*

Process and Data Flow

Having a good understanding of how chatbots work and what architecture will be involved, we can now dive into the specifics of what we want our chatbot to accomplish. Specifically, what intents will a user be able to invoke and what slots will be associated with each?

Storyboard/Sample Script

In order to come up with some great intents for our chatbot, it helps to do a little role-playing. For the sake of our chosen scenario, we will pretend we are a candidate exploring the site. After we have explored the career site for a while, the chatbot pops up on the screen. From here, we can make up a useful hypothetical dialogue between the candidate and the chatbot and then work backward to convert the conversation into specific intents and slots that define the scope of our chatbot functionality.

Chatbot: Hi there! Thank you for visiting our career site! Is there anything I can help you find?

Candidate: I'm looking for information on health insurance provided for company employees.

Chatbot: The company provides full health coverage to all full-time employees. For more information on benefits, the company benefits page can be accessed by clicking "benefits" in the top-level navigation. Is there anything else I can help you find?

Candidate: Thanks! I am also looking for Project Manager jobs.

Chatbot: We currently have two jobs posted that matches the search term "Project Manager." "Project Manager: HR Systems" and "Project Manager: Supply Chain." Would you like more information on these jobs or is there anything else I can help you find?

Candidate: None of these jobs interest me.

Chatbot: I'm sorry. Would you like me to alert you when more Project Manager jobs become available?

Candidate: Yes.

Chatbot: OK. To get started, please tell me your first name.

Candidate: Joe.

Chatbot: Thank you Joe. What is your last name?

Candidate: Schmoe.

Chatbot: OK Joe Schmoe, what is your preferred email address we can use to contact you about jobs?

Candidate: test1@test.com

Chatbot: Great! Would you also like to provide a phone number so we can contact you via text?

Candidate: Yes, it is 5555555555.

Chatbot: Thank you! Please know that your privacy is important to us. Our privacy policies are based on country of residence. May I please know your country of residence?

Candidate: USA.

Chatbot: Thank you! Please read the privacy agreement here <link> and let me know if you agree to the terms. By typing "Yes," you agree to the terms and I will create a job alert to your email address and phone number.

Candidate: Yes.

Chatbot: Thank you! Your job alert has been created! Is there anything else I can help you find?

Candidate: No, I'm good, thanks!

Chatbot: Thank you for visiting our career site. Goodbye!

Note   *The provided script is what is called a "happy path" script where no errors or terms the chatbot did not understand occurred. For the sake of brevity, we won't go through all of the negative testing a solid chatbot would require. However, in the event of an utterance the chatbot does not understand, you would want a set of statements prepared to help guide the user to something the chatbot does understand, a statement like "Sorry, I didn't understand that. You can ask me things like…"*

Defining Intents and Slots

With our sample script created, it looks like we have a few intents and corresponding slots defined. Let's start with the first interaction in the script. The candidate says they are "looking for information" on "health insurance." This can be analyzed into an intent of "content lookup" with a slot of <type of content>. This type of intent likely would not access SAP SuccessFactors to gather information. Instead, the chatbot would be configured to access a knowledge base in the event of a content lookup intent. The knowledge base would store what type of responses to give based on the type of content slot requested. Think of it as a table of prepared responses. Typically, this knowledge base would be stored in your chosen cloud chatbot solution. In this case, the chatbot script logic would be: in the event health insurance questions are asked, return the response "The company provides full health coverage to all full-time employees. For more information on benefits, the company benefits page can be accessed by clicking 'benefits' in the top-level navigation. Is there anything else I can help you find?"

In the next set of dialog, the candidate indicates they are "looking for… jobs." We can equate this to a "job search" intent with a slot <search term>, in this case "engineering." Since SAP SuccessFactors stores this information, we know we will have to have our chatbot query the system using the ODATA API to gather the response information.

Following the job search, the candidate indicates that they are unable to find their desired job. This triggers the "create profile and job alert" intent. This intent requires prompting the user to provide values for several slots: <first name>, <last name>, <email>, <mobile phone>, and <consent>. In addition, we already know the user is interested in engineering jobs, so we will also treat this as a <search term> slot.

Last, the candidate indicates they are finished chatting, which triggers the "log off" intent and requires no slots. We can see a visual representation of the intents and slots that define the scope of our chatbot as well as the subsequent scripting the chatbot executes as a result of each in Figure 5-3.

Note   *We have also included a <locale> slot to determine the language in which the chatbot and user are conducting intents for international implementations and because some objects within SAP SuccessFactors will require this information to query and create objects. Furthermore, <country> is required to determine which country consent statement will be shown to the candidate.*

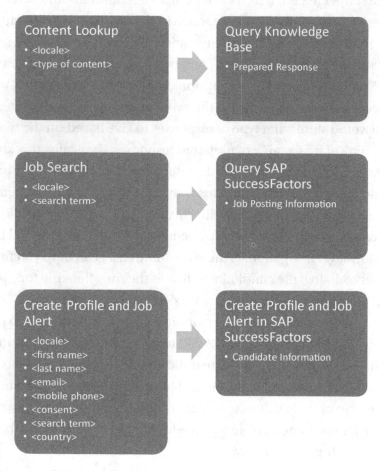

Figure 5-3. *Scope of Intents for Example Chatbot*

Relevant ODATA API Objects and Functions

Now that we've identified what specific intents and slots will be in scope for our chatbot, we can identify the APIs and objects these intents and slots will call in SAP SuccessFactors. Let's start with the job search intent. Similar to what we did in Chapter 2, we can use the Integration Center to explore the relevant ODATA objects and even start to build ODATA calls. Let's get started!

Job Search Intent

In order to gather the data needed to perform the job search intent and build the needed ODATA queries, follow the given steps:

1. Type and select "Integration Center" in the search bar.

2. The Integration Center main menu will appear. Click "My Integrations."

3. Click the "Create" button.

4. In the drop-down that appears, select "Scheduled Simple File Output Integration."

Note   *You may wonder, why am I creating a file output if I want to make an ODATA call to feed information to my chatbot? The answer is that we will show you a trick to save you time building ODATA calls from scratch by using this type of integration in the Integration Center. Keep reading to find out!*

---

5. Since we are interested in searching through active postings to accomplish our chatbot intent, we will start with the JobRequisitionPosting object. In the "Search for Entities by Entity Name" field, enter "JobRequisition" and click the first result on the left. The screen should appear as shown in Figure 5-4.

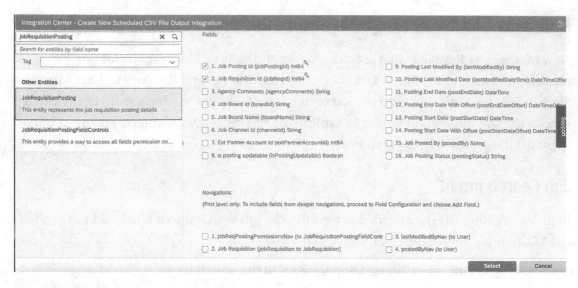

Figure 5-4. *JobRequisitionPosting Object Fields*

6. Our search will also need to filter for only those postings that are currently public and active on the external career site. Select the "Job Posting Status" and the "Job Board ID" fields and click the "Select" button.

7. The Options screen will show as seen in Figure 5-5. Choose "Simple Header" from the drop-down in the "Header Type" field. Click the "Next" button.

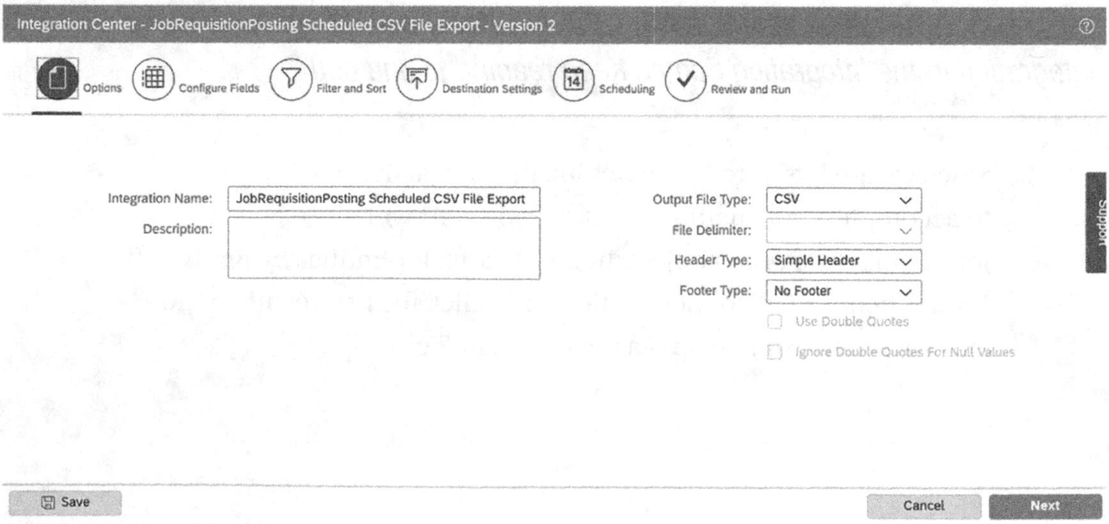

Figure 5-5. *Options Screen*

8. The Configure Fields screen will show. You will see a preview of the fields. You may notice a lot of postings for the internal career site or agencies, etc., or with a status value of expired. We want to make sure our chatbot only tells users about postings on the external career site that are active. Therefore, we will set some filters. Click the "Next" button to progress to the "Filter and Sort" screen.

9. Our first filter set will simply indicate that we only want to see postings for the external career site. Select "Job Board ID" for the Filter Set 1 "Field" column. Set "Operation" to "is equal to" and set the "Value" to "_external."

10. Click the "Add Filter Set" to add another set of filter parameters. This second filter will ensure only active postings are shown and not ones that are deleted or expired. Select "Job Posting Status" for the "Field" column and "is equal to" for the "Operation." Enter "Success" in the "Value" column. Click the "+" sign at the end of the row to add another parameter to the same filter set. Click the "Or" toggle beneath "Filter Set" 2. Enter "Job Posting Status" for the "Field" column, "is equal to" for "Operation," and enter "Updated" in the "Value" column. The result should look like Figure 5-6.

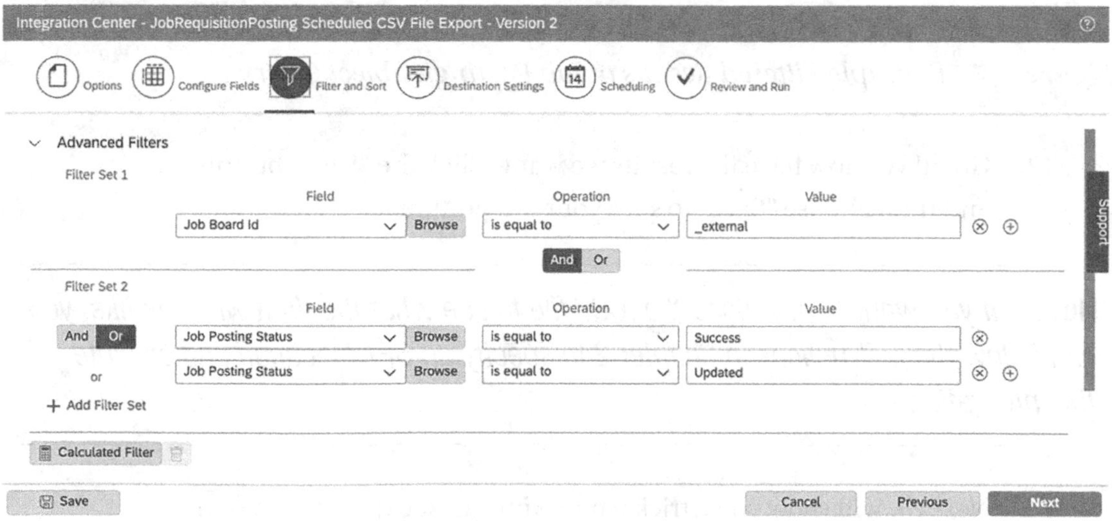

Figure 5-6. *Result of Entering Filter Parameters*

11. Click the "Previous" button to return to the "Configure Fields" screen. You should now only see results where the "Job Board ID" is equal to "_external" and the "Job Posting Status" is either "Updated" or "Success." An example is shown in Figure 5-7.

Note   *This assumes you have active, posted requisitions on the external career site in your SAP SuccessFactors system.*

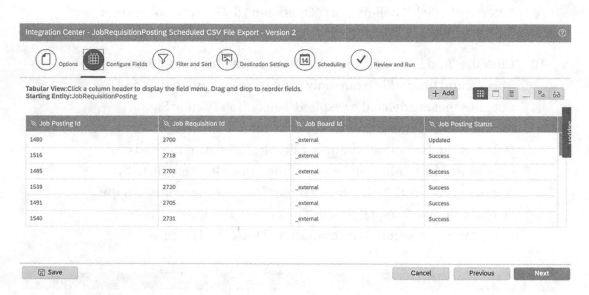

Figure 5-7. *Example Filtered Results of Job Postings Object Query*

12. Great! We now have the results we want. Click the "Save" button and then choose "Save" to save your integration.

Note   *If you want to download the .CSV file to see what the data will look like, you can follow the instructions in Chapter 2 to specify an SFTP location and execute the integration.*

---

13. Next we will show you a trick where you can see the ODATA query the integration uses behind the scenes to gather this data. This is extremely useful to use since chances are whatever cloud chatbot

solution you have chosen will need to make real-time ODATA calls and not rely on batch-driven .CSV files. Click the "Save" button again and then choose "Export Integration Specification."

14. This will download a .CSV file onto your local computer. Open the .CSV file using a text editor. An example of what this file will look like is shown in Figure 5-8.

```
Integration Name,JobRequisitionPosting Scheduled CSV File Export
Version,2
Description,null
Starting Entity,JobRequisitionPosting
Selected Fields,"jobPostingId,jobReqId,boardId,postingStatus"
Joined Entities,
Filter,boardId eq '_external' and (postingStatus eq 'Success' or postingStatus eq 'Updated')
Sort Order,

Format,CSV
Delimiter,
Header Type,SIMPLE
Footer Type,NONE

Schedule,The integration is not scheduled.
SFTP Host Address,
User Name,SFTPUser1
File Destination Folder,/
File Name,SampleJobPostings.csv


Field Number,Field Label,Description,SF Entity,SF Field,Default Value,Association,Data Type,Format,Value Lookup Table,Calculation,Maximum Length,Min

Body Fields
1,Job Posting Id,,JobRequisitionPosting,Job Posting Id (jobPostingId),,jobPostingId,Edm.Int64,,,,0,
2,Job Requisition Id,,JobRequisitionPosting,Job Requisition Id (jobReqId),,jobReqId,Edm.Int64,,,,0,
3,Job Board Id,,JobRequisitionPosting,Job Board Id (boardId),,boardId,Edm.String,,,,0,
4,Job Posting Status,,JobRequisitionPosting,Job Posting Status (postingStatus),,postingStatus,Edm.String,,,,0,


OData Query URL,,"/odata/v2/JobRequisitionPosting?$select=jobPostingId,jobReqId,boardId,postingStatus&$filter=boardId eq '_external' and (postingSt
```

Figure 5-8. *Sample Integration Specification Document*

15. You will notice the last line of the specification contains the query URL! An example is shown as follows:

```
"/odata/v2/JobRequisitionPosting?$select=jobPostingId,jobReq
Id,boardId,postingStatus&$filter=boardId eq '_external' and
(postingStatus eq 'Success' or postingStatus eq 'Updated')"
```

Note   *For more information on making ODATA query calls, please refer to Chapter 1.*

We should now have the information we need to feed our chatbot a list of requisitions that have active postings on the external career site. Now, we need to query the JobRequisition object for those specific requisition IDs and conduct a search for our search parameter in order to see what the title is for each of those postings. Follow the given steps to build the needed query:

1. Type and select "Integration Center" in the search bar.

2. The Integration Center main menu will appear. Click "My Integrations."

3. Click the "Create" button.

4. In the drop-down that appears, select "Scheduled Simple File Output Integration."

5. We want to search through the posted content of the requisition for the candidate's specific language they are using to browse the career site. This information is stored in the JobRequisitionLocale object. In the "Search for Entities by Entity Name" field, enter "JobRequisitionLocale."

6. Choose the first option that appears below, "JobRequisitionLocale."

7. The screen will update on the right with a listing of all of the fields and navigation within the JobRequisitionLocale Entity. Select the fields "External Job Description," "External Title," "Locale," and the navigation to "JobRequisition." An example is shown in Figure 5-9. Click "Select."

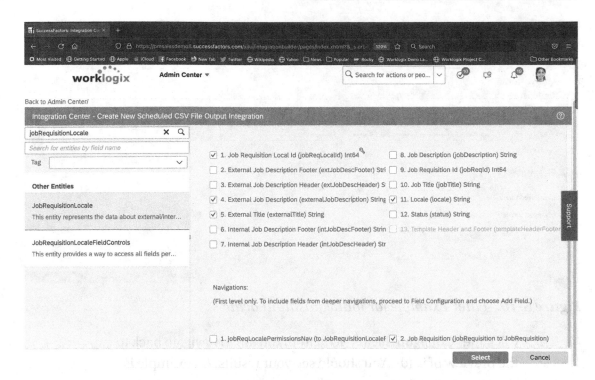

Figure 5-9. *JobRequisitionLocale Fields*

8. The Options screen will appear. Click the "Filter and Sort" icon toward the top of the screen. We will only want to look through the list of active jobs we received in the last query that are also in the user's language. In our example, we will just filter one job requisition ID and use en_US (the locale for United States English). Under Advanced Filters, select "Locale" under the "Field" column, "is equal to" under the "Operation" column, and enter the locale of your posting under the "Value" column (in our example, we use "en_US"). Click the "+" icon to the right of the row. Select "Job Requisition Id-JobRequisition" under the "Field" column, "is equal to" under the "Operation" column, and enter one of the requisition numbers from the previous query we build in the "Value" column (in our example, we use "2700"). Repeat as needed for all the requisitions returned in the last query. An example is shown in Figure 5-10.

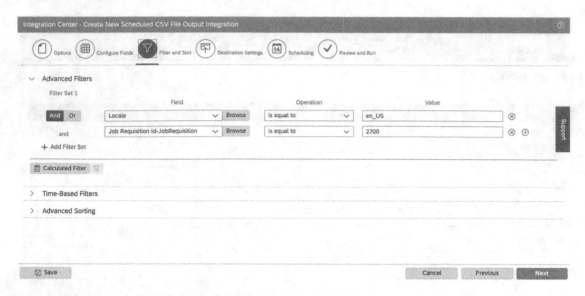

Figure 5-10. *Filter Example for JobRequisitionLocale*

9. Click the "Configure Fields" icon from the top to navigate back to the preview of fields. You should see your results. An example is shown in Figure 5-11.

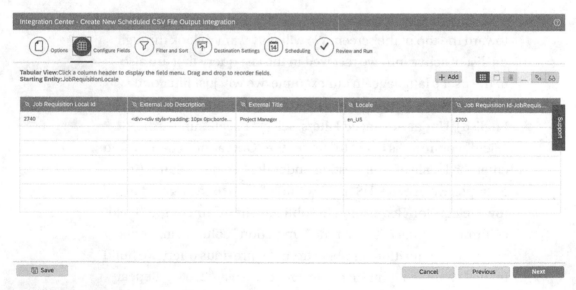

Figure 5-11. *Example Results of JobRequisitionLocale Filters*

10. Click the "Save" button in the bottom left-hand corner, and in the menu that appears, click "Save." In the pop-up that appears, click "Save."

11. Once again, we will want to grab the ODATA URL call in the integration specification file. To download the integration specification, click "Save" and then click "Export Integration Specification."

12. Open the file with your text editor. You should see the generated ODATA call URL. An example is shown as follows:

```
/odata/v2/JobRequisitionLocale?$select=jobReqLocalId,
externalJobDescription,externalTitle,locale,
jobRequisition/jobReqId&$expand=jobRequisition&$filter=
(locale eq 'en_US' and jobRequisition/jobReqId eq 2700)
```

You now have the second piece of the puzzle that allows your chatbot access to the content of the job postings and relay it to the candidate on the career site! With this, we have all of the information from SAP SuccessFactors to complete our job search intent! In the next section, we will look at the create profile and job alert intent.

Note   *You can sort through this content using your own search patterns for the search term (in our example, "Project Manager"), or you could use the ODATA $search parameter. For more information see, the API specifications from SAP here:* `https://api.sap.com/api/RCMJobRequisition/resource.`

Create Profile and Job Alert Intent

The second intent in scope for our chatbot that will require an ODATA interaction with SAP SuccessFactors is the create profile and job alert intent. In this intent, the chatbot gathers the user's first name, last name, email address, mobile phone number, and their consent to create an ID and be contacted about future job opportunities relating to their search term. This involves making an ODATA call to create a candidate profile. Similar to what we did in Chapter 2, we will walk through the Integration Center to see what an example of candidate data looks like in our system and then create a call to create data in the proper format. Remember, it is best practice to know how the system will want the data formatted for and what fields are available in your specific implementation before attempting to create an object. Also remember that fields can vary from customer instance to instance based on business requirements and configurations made in the candidate XML. Follow the given steps to walk through the Candidate object data and create an ODATA call to create a candidate profile:

1. Type and select "Integration Center" in the search bar.

2. The Integration Center main menu will appear. Click "My Integrations."

3. Click the "Create" button.

4. In the drop-down that appears, select "Scheduled Simple File Output Integration."

5. In the "Search for Entities by Entity Name" field, enter "Candidate."

6. Choose the first option that appears below, "Candidate."

7. The screen will update on the right with a listing of all of the fields and navigation within the Candidate entity. Select the fields "Candidate ID," "Agreed to Privacy Statement," "Candidate Locale," "Mobile Number," "Consent To Marketing," "Primary Email," "Country of Residence," "First Name," "Last Name," and "Share Profile flag." An example is shown in Figure 5-12.

Note   *We are including fields "Agreed to Privacy Statement," "Consent To Marketing," "Country of Residence," and "Share Profile flag" in order to properly document that the candidate has agreed to the terms shared by the chatbot via a link. By default, we would think that you would want candidates to consent to marketing so your recruiters can send them marketing based on their search term (controlled by the "Consent to Marketing" field) and to be searchable by all recruiters in your system (controlled by the "Share Profile flag" field). Different countries have different requirements for acquiring consent to store personal information and use it for marketing purposes. Thus, we capture the "Country of Residence" field to know which privacy consent form they have agreed to and document their agreement in the "Agreed to Privacy Statement" and Privacy Acceptance Date" fields. You should consult with your compliance team to avoid* *violating any privacy laws. We also recommend always including the candidate locale when creating a candidate to avoid errors with downstream processes. For example, whenever an application is created, it looks to the candidate locale to determine the locale of the application.*

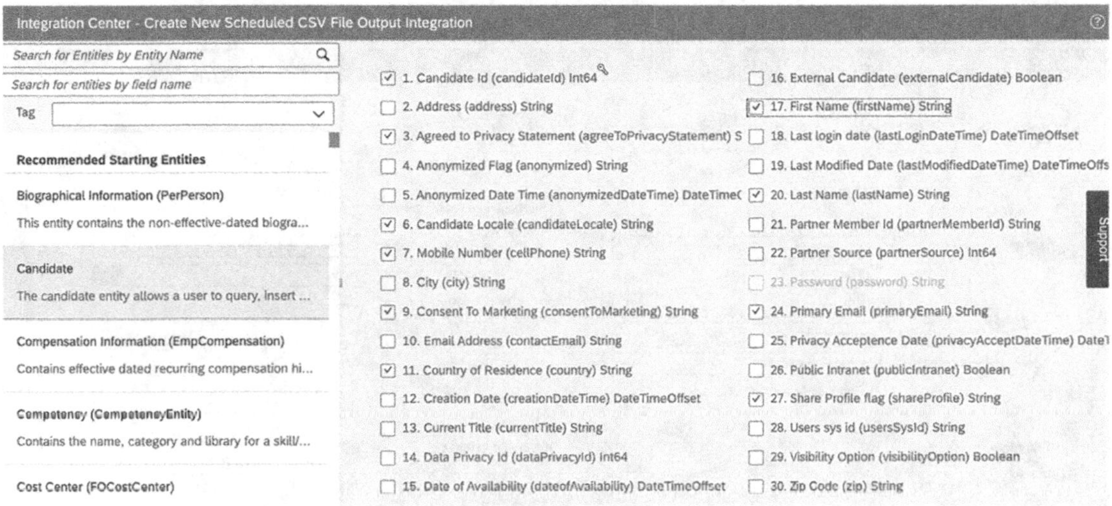

Figure 5-12. *Selected Relevant Fields for Chatbot Creation of Candidate Profile*

8. You will be taken to the Options screen. Select "Simple Header" in the Header Type field. An example is shown in Figure 5-13.

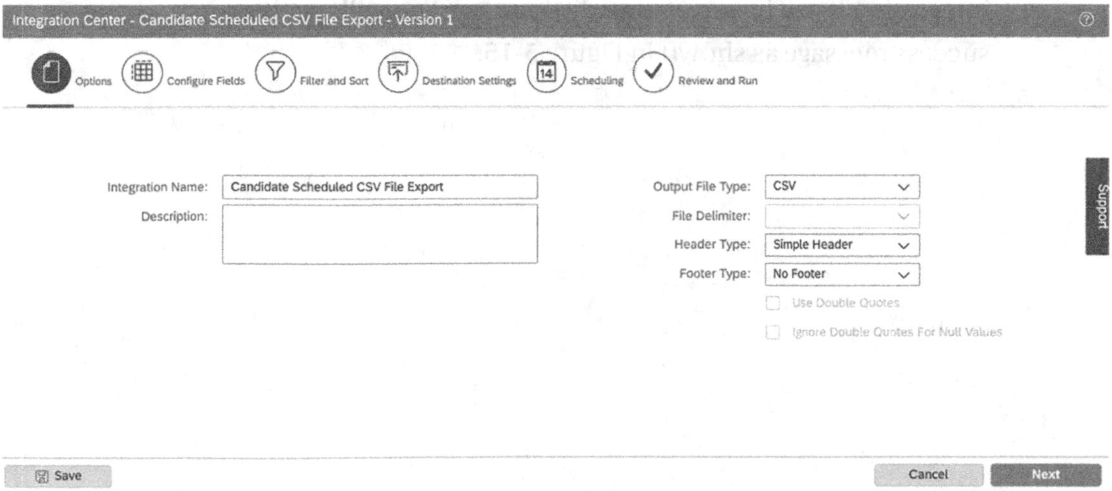

Figure 5-13. *Options Screen for Candidate Object CSV File Export Integration*

119

9.   We will want to export the .CSV file so we can edit the file to use
     for import. Click the "Destination Settings" icon toward the top
     of the screen, and fill in your STFP host address, username, and
     password information. Be sure to give the file a name and the
     .CSV extension. An example is shown in Figure 5-14 (minus the
     SFTP site).

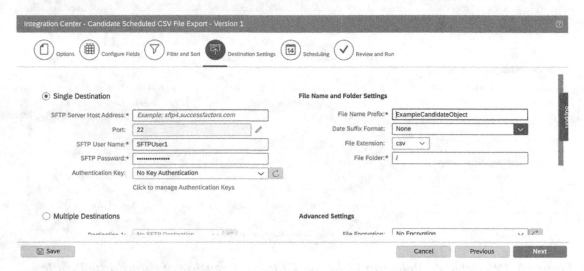

Figure 5-14. *SFTP Destination Settings for Example Candidate Object Export*

10. Click the "Review and Run" icon toward the top of the screen.
     And then click "Run Now" (you may be prompted to save your
     integration if you have not already done so). You will receive a
     success message as shown in Figure 5-15.

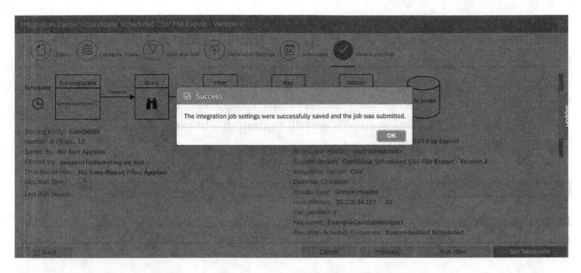

Figure 5-15. *Success Message for Integration Run Submission*

11.   Click the refresh icon next to "Last Run Time" periodically to see
       if the integration has completed running successfully. Once it is
       finished, check your SFTP site and download the generated file.

Note   *Alternatively, if you only have a limited amount of data and/or no SFTP
server available, from the "Configure Fields" screen, you can click the glasses
icon in the upper-right-hand corner of the screen to view the file on screen. An
example is shown in Figure 5-16.*

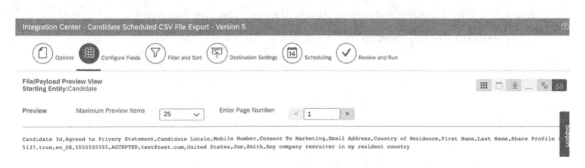

Figure 5-16. *Preview of .CSV File Online*

Similar to what we did in Chapter 2, you will now want to open the .CSV file and edit it. Remove unneeded rows so that only a good sample candidate is left. Remove the "Candidate ID" column and the corresponding value. You will also want to create a new test email address since that is the unique identifier checked when creating a new Candidate object. Also, change the value of "Share Profile flag" to "0". This will help you avoid an error in the upcoming steps – we will explain this in more detail later. Example .CSV file content is shown in Figure 5-17.

```
Candidate Locale,Agreed to Privacy Statement,Mobile Number,Consent To Marketing,Primary Email,Country of Residence,First Name,Last Name,Share Profile flag
en_US,true,5555555555,ACCEPTED,test1@test.com,United States,Joe,Schmoe,0
```

Figure 5-17.   *Example .CSV File Used to Construct and Create Candidate Integration*

Now that we've got some good sample data, let's build an integration that creates a candidate! Once again, we can use the Integration Center to build this out and then grab the resulting ODATA call! Follow the given steps to create the ODATA call:

1. Type and select "Integration Center" in the search bar.

2. The Integration Center main menu will appear. Click "My Integrations."

3. Click the "Create" button.

4. Choose "Scheduled CSV Input Integration."

5. In the "Search for Entities by Entity Name" field, type "Candidate."

6. Click the "Candidate" object that will appear as the first search result on the left as seen in Figure 5-18. Then click "Select."

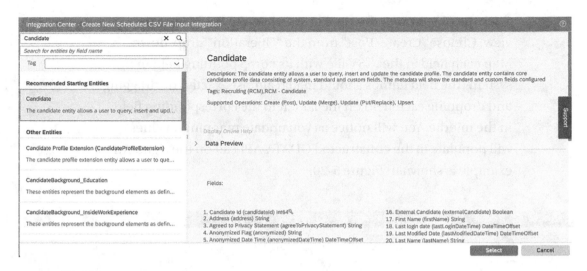

Figure 5-18. *Selecting Candidate Object for Input*

7. Enter a name for your integration and then click the "Next" button. The "Configure Fields" screen will appear.

8. Click the "Upload Sample CSV" button. Choose the .CSV file you created in the prior steps. An example of the final result is shown in Figure 5-19.

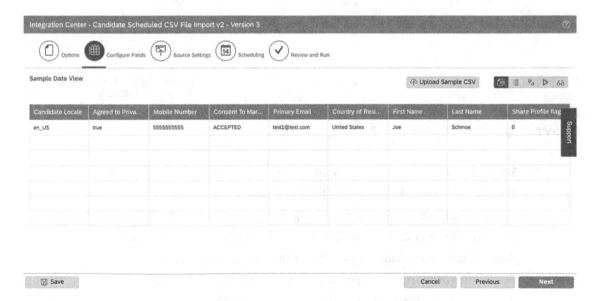

Figure 5-19. *Example .CSV File Loaded into Integration Center for Candidate Creation Integration*

9. Click the plug icon to the right to switch to the field mapping
    view. Choose "Create/Post" from the "Operation" drop-down.
    Map each field in the .CSV file with its corresponding field in the
    system (the field names should match almost exactly) by dragging
    and dropping each field on the far left to the corresponding field
    in the middle. You will notice as you update mappings, values
    will populate in the constructed ODATA API call on the right. An
    example is shown in Figure 5-20.

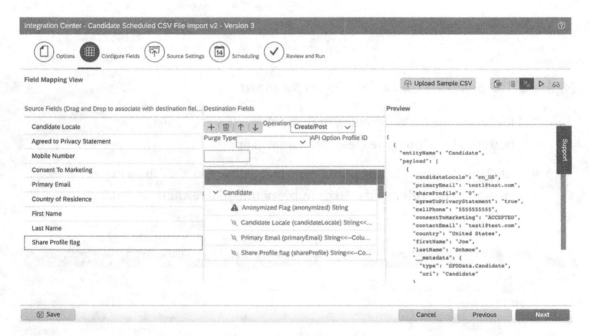

Figure 5-20. *Example Field Mapping and ODATA Call for Creating a Candidate
Object*

10. You will notice that there is a caution sign next to the "Anonymized
    Flag" field. This is because the field is required for a create
    operation. Since we are just creating the candidate for the first
    time, we do not need to anonymize the information (this is only
    done when a candidate is purged from the system and sensitive
    fields are anonymized to comply with privacy laws). Therefore,
    we will default the field to "false." To do this, click the bulleted list
    icon in the upper-right-hand corner of the screen. This will take

you to the "Field Detail" view. Click the "Anonymized Flag" field on the left. In the "Default Value" field, type "false." An example is shown in Figure 5-21.

Figure 5-21. Setting Default Value for "Anonymized Flag" field

11. Click the glasses icon in the upper-right-hand corner of the screen to see the updated content of the constructed ODATA call. An example is shown as follows:

```
[
  {
    "entityName": "Candidate",
    "payload": [
      {
        "anonymized": "false",
        "candidateLocale": "en_US",
        "primaryEmail": "test1@test.com",
        "shareProfile": "0",
        "agreeToPrivacyStatement": "true",
        "cellPhone": "5555555555",
```

```
          "consentToMarketing": "ACCEPTED",
          "contactEmail": "test1@test.com",
          "country": "United States",
          "firstName": "Joe",
          "lastName": "Schmoe",
          "__metadata": {
            "type": "SFOData.Candidate",
            "uri": "Candidate"
          }
        }
      ]
    }
]
```

12. Next, we can test our ODATA call. Click the play icon in the upper-right-hand corner of the screen. Then click the "Run Preview Records" button. The screen will update to show the status of the run in the "Run Results Status" column. If the run is successful, a checkbox will appear in the column. An example is shown in Figure 5-22.

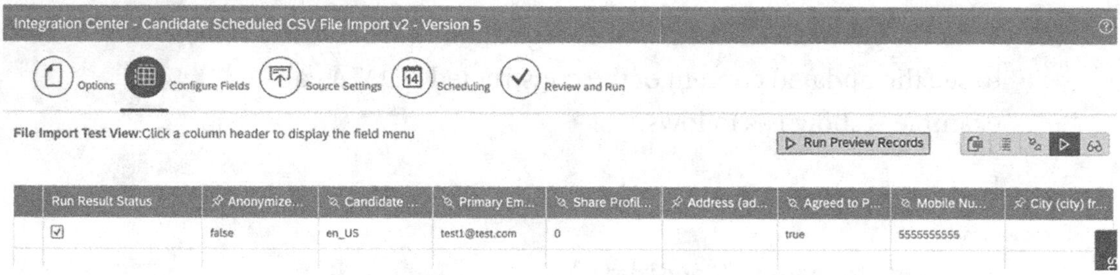

Figure 5-22. *Successful Creation of a New Candidate Object*

Note   *Every now and again, the data output by an Integration Center integration is not exactly in the format SAP SuccessFactors requires for input (Please refer to SAP note 2874787 for an explanation of the shareProfile tag.). Such is the case with the "shareProfile" field we had you change in your CSV file earlier. While a long string value is what is output, the input required is a corresponding code with* *each value. Luckily, we can see in the error messages what the expected value is, in this case a "0", "1", or "2". Through trial and error or loading test values, we can figure out that a "0" corresponds to our desired value of "Any company recruiter" and use the "0" value in our call. An example is shown in Figure 5-23.*

Figure 5-23. *Error Shown If "Any company recruiter" Value Is Not Changed to "0"*

Congratulations! You have created a Candidate object and now have an example ODATA call you can now use to call the SAP SuccessFactors system from your cloud chatbot solution and create more candidates!

There is also one slot from our create profile and job alert intent field that we know about that we have not accounted for yet: search term. In our examples thus far, we have only looked at standard fields. Since there is no standard field or API to create the standard job alert, our recommendation is to store this in a custom field. A custom field can be created on the Candidate or CandidateProfileExtension object to store this term. The term could then be used to organize candidates into talent pools and use the standard email campaigns functionality to reach out to candidates based on the value stored in the custom field. Alternatively, another custom chatbot could reach out periodically to these talent pools.

Data Flow Across Intents and Slots

Now that we have dug into the specifics of what fields and APIs exist in SAP SuccessFactors to support our chatbot scope, we can take a step back and construct a diagram that shows how data would flow to and from SAP SuccessFactors and the cloud chatbot solution. Our diagram is shown in Figure 5-24.

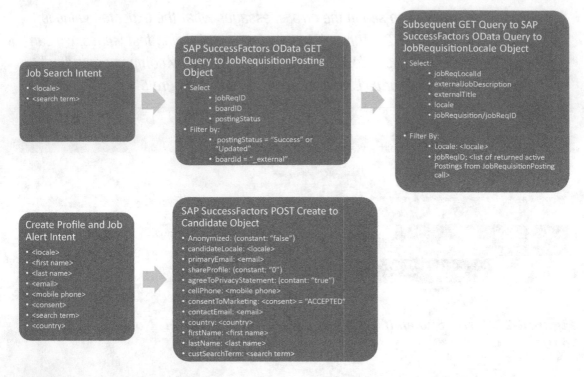

Figure 5-24. *Data Flow Between Cloud Chatbot Solution and SAP SuccessFactors*

Case Study End Results

We now have the tools in SAP SuccessFactors ready to go to build our chatbot in whatever our chosen chatbot platform will be. For example, we now know how to query job postings and create a candidate profile! Having this profile allows us to engage with the candidate even more. Recruiters are able to look up the candidate information and match the candidate to job postings using the standard SAP SuccessFactors candidate search functionality. Furthermore, if we have built other automations and enhancements such as chatbots or emails that alert candidates to upcoming career fairs or job postings, this candidate's information is now available for those automations and enhancements to engage.

Conclusions

Congratulations! You have now examined different strategies and tools for increasing candidate engagement throughout the candidate experience in SAP SuccessFactors. In addition, you have walked through a practical example of how to use SAP SuccessFactors ODATA APIs to create a chatbot that assists candidates on your company career site and increases candidate engagement. At this point, you should be able to understand how some of these other strategies laid out in the first section could be implemented as well. We hope you have sparked your creativity! Using this same method of designing and creating the chatbot for the career site, we hope you will be able to create your own unique candidate experience enhancements and automations across the candidate experience process. One of the most exciting aspects of SAP SuccessFactors is that it serves as a solid platform where many enhancements can be built and work together to enhance the candidate experience.

CHAPTER 6

Robotic Process Automation

In this chapter, we will cover the basics of Robotic Process Automation (RPA) and how you can use the techniques to improve or enhance your SAP SuccessFactors Recruiting implementations.

Introduction

As software systems have evolved and become more complex over time, there's a need to find ways to operate systems cheaply and at scale. Robotic Process Automation is one such approach. What is Robotic Process Automation (RPA)? Wikipedia defines this as "a form of business process automation technology based on metaphorical software robots (bots) or on artificial intelligence (AI) digital workers. It is sometimes referred to as software robotics (not to be confused with robot software)."

RPA streamlines workflows, which makes organizations more profitable, flexible, agile, and responsive. The notion is to relieve the workers from performing routine mundane tasks and thus increase employee satisfaction, engagement, and productivity. As a form of automation, it consisted in screen scraping, also unfortunately a source of early forms of malware. The barrier to adoption of RPA into existing software systems is often technological: it's not always economically viable to retrofit new interfaces into existing systems. RPA software provides a pragmatic means of deploying new services in such situations, where the robots simply mimic the behavior of humans as they would interact with the software. Where this approach becomes useful is that no new information technology transformation or investment is required.

© Anand 'Andy' Athanur, Mark Ingram and Michael A. Wellens 2022
A. A. Athanur et al., *Innovative SAP SuccessFactors Recruiting*, https://doi.org/10.1007/978-1-4842-7425-5_6

Vendor Choices

Naturally, several vendors have emerged in this area and hope to live up to the promise of RPA, namely, to make automation cheaper and repeatable. You can search for different vendors by starting with `https://research.aimultiple.com/robotic-process-automation-rpa-vendors-comparison/`. There are a few pure-play RPA vendors in this rapidly expanding marketplace. Examples include UI Path, Automation Anywhere, Blue Prism, Workfusion, and Pegassystems. Some enterprise vendors have their own solutions as a natural extension to their existing enterprise software or partner with pure-play RPA vendors. Examples include SAP's own iRPA and Oracle Integration Cloud. There are a few open source RPA alternatives that can be implemented without any licensing costs. Examples include Robot Framework RPA, openRPA, UI.Vision, Robocorp, Taskt, and TagUI. We suggest that you start small with the open source versions and, then as you gain more expertise, expand into licensed versions. In this manner, you will keep your costs low and expand as your needs change.

Choosing an RPA solution will depend on the depth of the product road map, user interface, whether you want an attended or unattended robot, and finally the level of AI and ML on the RPA platform.

Relevant Business Scenarios

RPAs have found several uses in the industry at large depending on the specific problem they're trying to solve. Without going into the specifics, their applications include customer service, invoice processing, sales orders, payroll, price comparison, storing customer data, processing HR information, processing fast refunds, recruitment, and extracting data from different formats in combination with optical character recognition (OCR).

Pros and Cons of Employing RPA

Like every other new technology/approach, we need to be aware of the promise and the practical challenges of using RPAs.

Pros:

- RPAs are relatively cheap and there are plenty of tools available, licensed, and free.

- Some RPAs require little or no technical skills to implement. However, the depth of domain knowledge will be the deciding factor in the success of the implementation.

- Increasingly, more customers are aware of RPA tools and the promise they hold in terms of automating repetitive tasks.

- They're easier to implement than API-based solutions; however, their scope can be limited due to them relying on UI components.

- In some cases, the only requirement is to understand the UI components without paying attention to the underlying infrastructure.

- If the UI is well understood and the execution path of the automation is predictable, RPAs will prove to be a good return on investment.

Cons:

- While they're relatively cheap to implement, APIs are inherently more powerful in terms of what can be done and achieved.

- RPA robots require that the implementor understand the UI flow and execution paths quite well and will require the input of customers' users to be truly effective.

- RPAs will be a security challenge for many customers because they may be unwilling to expose usernames and passwords to external entities.

- Since RPAs rely on UI behavior, they may be a little more complex to implement for multi-locale implementations.

- If the UI for the underlying solutions change frequently, RPAs will need to be updated and maintained on an ongoing basis.

- If the UI has interactive behavior, for example, if the UI changes based on the selection of a drop-down value, the RPA implementation can become quite complex and challenging.

- Scope creep will always be a challenge when you try to implement an automation solution.

When to Use RPAs for Recruiting?

In this section, we will cover some use cases for RPAs in SuccessFactors Recruiting. Recruiting has built in automation to deal with several tasks such as duplicating job requisitions, populating certain job requisition fields based on the position chosen, or even creating a job requisition from within SuccessFactors Employee Central. In addition to in-built automation, SuccessFactors Recruiting also has several ODATA APIs. Chapters 1 and 9 of this book deal with different aspects of ODATA APIs, their use, and application.

By its very nature, SuccessFactors Recruiting is quite a large system with a lot of technical aspects and business functionality. As described in the introduction of this chapter, RPAs can have several uses, but the decision on when to use RPAs is nontrivial. To successfully implement RPAs for a customer's recruiting implementation, here are a few suggested questions to ask the customer:

- Are they currently facing challenges with recruiting tasks that occupy a lot of recruiters' time and efforts?

- How complex is their recruiting implementation in terms of recruiting templates, number of fields that need to be supplied for a successful job requisition, and the associated applicant workflows (also known as applicant status workflows)?

- Are they currently automating tasks within their recruiting implementation?

- Do they have the technical infrastructure and/or middleware to aid in their automation?

- Are there any APIs available for the kind of automation they wish to accomplish?

- Are they outsourcing any recruiting tasks? In some cases, an RPA may mitigate the need for outsourcing.

- Do they have the resources available to build API-based integrations?

- Is the customer's recruiting centralized or distributed? The answer will drive your RPA efforts.

- Do they want attended RPA or unattended RPA? The answer will drive your choice of RPA tool.

- Can the customer automate their recruiting implementation by using business rules? Chapter 11 covers the practical implementation of business rules.

The list of questions is by no means exhaustive or comprehensive. Consulting with the customer will at least get you on the path to understanding a customer's challenges and investigate if an RPA approach will address the customer's needs. The intent of any automation or indeed RPA is to drive efficiencies and keep cost of ownership low.

Use Cases for Robotic Process Automation in Recruiting

Given what we have described in this chapter, it's probably a good point to ask the question, what are the possible use cases for RPA in recruiting? It's nearly impossible to come up with a comprehensive list of use cases because it will depend on the specific customer situation. For some customers, using APIs to achieve automation may be challenging due to resource constraints both in terms of technical resources and budgets. The same problem may be easy for one customer to address while it might be a challenge for another customer. What might be a challenge for large customer (because of the complexity of the implementation) might be easily solvable for a smaller customer.

Describing use cases for recruiting will be largely dependent on prior experience with the recruiting implementation and the current customer's implementation. Let's look at a few use cases for RPAs:

1. Automatically setting an assessment package for a job requisition

If you're creating a job requisition within SuccessFactors, you're required to provide mandatory fields; otherwise, you will not be able to complete the job requisition. While this is expected behavior, adding an assessment to a job requisition is, by definition, not required. Many organizations find this addition of an assessment to a job requisition a time-consuming activity. Unfortunately, at the time of this writing, there's no programmatic manner or API to set the assessment at the time of creating the job requisition. Figure 6-1 shows an example of a job requisition where the assessment must be added.

Figure 6-1. *Adding an Assessment to a Job Requisition*

Chapter 8 goes into greater detail on how to set up an assessment integration. For the purposes of this chapter, it's enough to understand that an assessment must be added to the list of job requisitions that the customer will provide. To be certain, there is some amount of automation that SuccessFactors Recruiting system provides you. For example, Figure 6-2 shows how you can duplicate a requisition and make multiple copies.

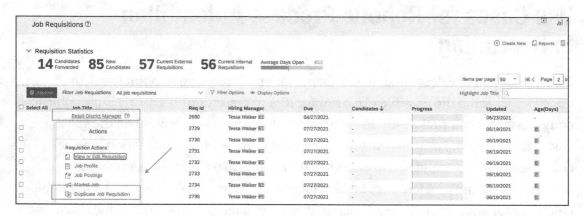

Figure 6-2. *How to Duplicate a Job Requisition*

Figure 6-3 shows a screenshot asking you to confirm how many copies you want to make.

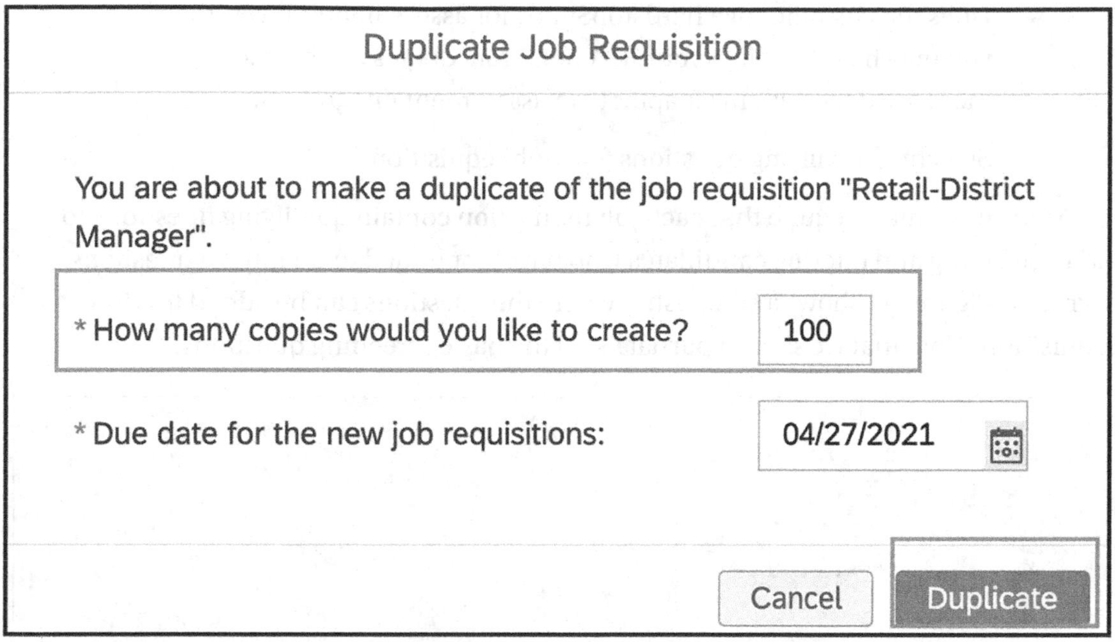

Figure 6-3. *Pop-Up Screenshot to Show How Many Copies of a Requisition You Want to Make*

Assuming that you have multiple copies of replicated job requisitions, the next step is to use an RPA bot to associate an assessment with the requisition. Once your RPA bot is set up, the next logical step is to have the bot automatically associate an assessment. Figure 6-1 shows a screenshot where the assessment should be added. Given this situation, you should be aware of the customer's configuration to make meaningful automation possible. The following list has a few questions that you need to consult with the customer:

- How many assessment vendors are there in the customer instance? If the answer is only one vendor, your solution is that much easier.

- How many unique combinations of assessment vendors and packages are possible? Depending on the situation, you may have to set up as many versions of the bot as there are unique combinations of assessment vendors and packages. It's a good idea to name your bots uniquely for each combination.

- Does the customer use separate email templates for a vendor/package combination? Depending on the answer, you may have to adjust your bot setup.

137

> – Does the customer use hardstopStatus for assessments? If yes, then
> you may have to adjust your RPA bot. More details on this require-
> ment are dealt with in Chapter 8 on assessment integration.

2. Selecting qualifying questions for a job requisition

Many customers require that each job requisition contain qualifying questions to aid in selecting and filtering candidates who have responded to qualifying questions correctly. Figure 6-4 shows a screenshot where the questions can be added to a job requisition. Note that we show a partial list of the basic screening questions.

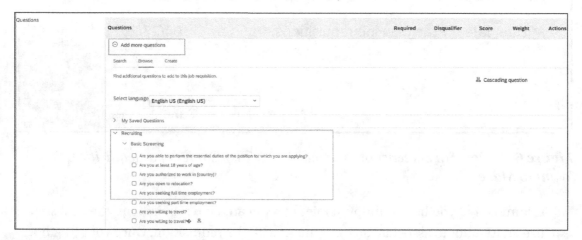

Figure 6-4. *Screenshot Showing the Initial Screening Questions for a Job Requisition*

Figure 6-5 shows a sample job requisition with some questions added including disqualifying questions.

Questions	Required	Disqualifier	Score	Weight	Actions
Are you at least 18 years of age? Multiple Choice	☑	☑	☐	0.0	Select ∨
Are you authorized to work in [country]? Multiple Choice	☐	☐	☐	0.0	Select ∨
Are you willing to travel?◆ Multiple Choice	☐	☐	☐	0.0	Select ∨
⊕ Add more questions					

Figure 6-5. *Sample Questions Added Manually to a Job Requisition*

This process of adding qualifying questions to a job requisition looks simple; it becomes time consuming and repetitive if you have to follow the same steps for 100s or even 1000s of job requisitions. In such cases, it's worthwhile considering an RPA bot solution. However, to have an efficient solution, you must consult with the customer on what the exact requirement is. Here are a few sample questions to consider before arriving at a final solution:

- Has the customer considered using a metadata framework (MDF) solution that has been proposed in Chapter 3? While a different solution using the metadata framework (MDF) has been proposed for addressing this issue, it certainly requires some understanding of MDF and Integration Center.

- Are the qualifying questions the same for a large number of requisitions or a subset of requisitions? Depending on the answer, you may have to tweak the RPA bot to address different sets of questions for different kinds of job requisitions.

- Is the recruiter who creates the job requisition in different locales (en_US, fr_CA, de_DE, etc.) or even different geo-location? Depending on the answer to this question, you must adjust your RPA to address this situation. The more locales you must account for, the more complicated your RPA bot will have to become. It's very likely that you need an RPA for each locale.

3. Re-triggering an assessment for a candidate

   This is not a common use case, but it is possible especially when the customer has lots of data (job requisitions, job applications, and candidates). If you read Chapter 8 on managing assessments, there may be an uncommon situation where the vendor's system is unavailable for some reason. In such a situation, a few 100s or even a few 1000s of job applications have the assessment triggered within SuccessFactors, but due to the unavailability of the vendor's system, SuccessFactors candidate summary page will display a message: "Error: unable to initiate assessment." The simple workaround when the vendor's system is available again is to reinitiate the assessment in the UI. However, as explained here, the issue is the number of assessments that need to be initiated manually. The challenge is that at the time of this writing, there's no API solution to reinitiate the assessment. In such a situation, an RPA bot to reinitiate the assessments makes sense. Like every use case for an RPA bot, you must discuss the scope of the solution. Here are a couple of suggested discussion points to consider:

- Are the assessments tied to one job requisition or are they spread across several job requisitions? The answer to this question will determine the execution path for the RPA.

- Another option is for the vendor itself to re-trigger the notification to the candidate to undergo the assessment. Since the vendor already knows which assessments were triggered from SuccessFactors but failed, it's a matter of knowing how many assessments failed. The vendor can then query the JobApplicationAssessmentOrder API to gather the details about the candidate, job requisition, and the job application.

Figure 6-6 shows a screenshot where the assessment can be reinitiated en masse on the candidate summary. To do this, select one or more candidates and click the drop-down against "Action" label. However, in this screenshot, the label for "Initiate Assessment" is unavailable since these sample candidates have already undergone assessment.

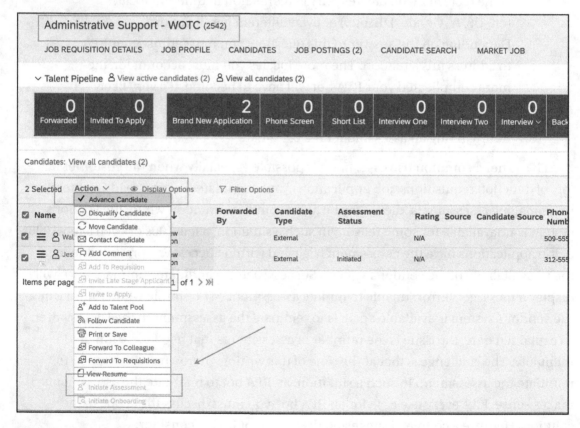

Figure 6-6. *Screenshot of Candidate Summary Page Where Assessments Can Be Mass Initiated*

4. Advancing a candidate in the UI

Figure 6-6 shows that one of the options available in the Action's drop-down is to Advance Candidate for one or more candidates. Advancing a candidate means moving the candidate one step forward in the applicant workflow. In most configurations, advancing a candidate one step at a time is often not required; however, there may be certain configurations where a step is required and cannot be skipped without providing a justification by the user. In such cases, an RPA bot can come in handy. Please note that this requirement is entirely dependent on the configuration.

Here are a few questions to consult with the customer:

– Does the customer have required steps in the applicant workflow, and does the recruiting user need to provide a justification for skipping a status? If yes, then it's worthwhile to ask the customer what the nature of the justification is. In some cases, it might simply be a checkbox that needs to be clicked or a comment provided. Your RPA bot should account for this checkbox or comment.

– Does the customer require that each candidate be advanced one step at a time? Your RPA bot should then ideally be configured to account for each status change. It's likely that you will end up with as many bots as there are statuses to be advanced with minor changes at each status. A reminder at this stage is to keep in mind that the aim of the RPA bot should be to automate tasks at a large scale and not get bogged down by having to design a bot for a handful of cases. In cases where there are few instances, it might be simpler to do those tasks manually.

5. Moving a candidate to a specified step in the UI

This is a variation on the previous step. In this situation, the candidate is advanced to a specified step in the applicant workflow. Of course, it's dependent on the customer's recruiting configuration. Figure 6-7 shows an example of moving two or more candidates to a specified step in the workflow.

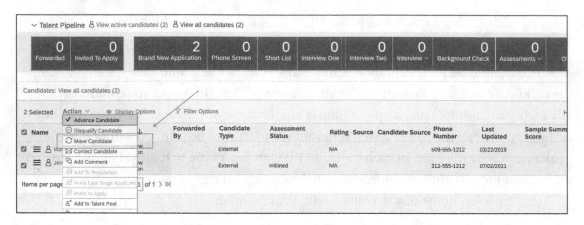

Figure 6-7. *Screenshot Showing How to Move a Candidate*

Figure 6-8 shows the pop-up acknowledgment that the system displays when moving one or more candidates.

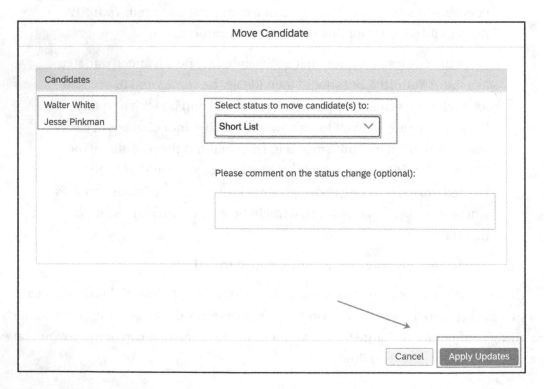

Figure 6-8. *Acknowledgment Screen Requesting Confirmation of Selected Status*

This might work in most situations; however, the Move Candidate action will not work for the following reasons:

1. A hardstopStatus may be configured on the assessment that will prevent moving the candidate(s) to a desired stop. Please refer to Chapter 8 on assessment integrations in this book for more information.

2. The customer may have unskippable statuses with required fields. In such cases, the only option is to move the candidate(s) to the last skippable status in the applicant workflow.

Figure 6-9 shows another variant of Figure 6-10 where a particular status may have sub-statuses or subitems. The option displayed is dependent on the individual customer's configuration.

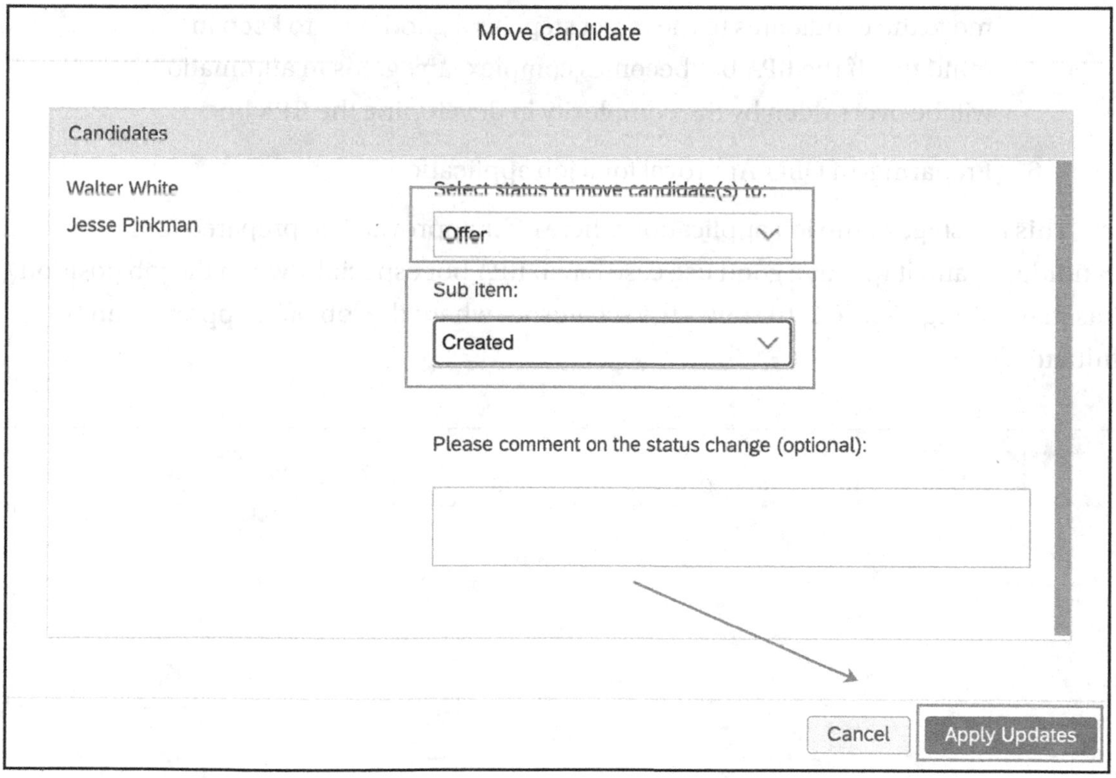

Figure 6-9. *Acknowledgment Screen Showing a Status with Subitem*

To address this RPA situation, you need to consult with the customer. Here are a few questions to consider:

- Does the customer have one or more assessments configured in their tenant, and have they defined a hardstopStatus for their assessments? Keep in mind that the hardstopStatus is configured at the job requisition level and will vary within a customer. This will impact the way you define the RPA bot.

- Does the customer have unskippable statuses? If there are unskippable statuses, your RPA bot must accommodate for this requirement.

- Does the customer have required fields before the target status? If that's the situation, you may have to modify the RPA to first move to the step that has required fields, fill in the required fields, and finally move the candidates to the target step. It's a good time to keep in mind that if the RPA bot becomes complex, any gains in automation will be overridden by the complexity in developing the RPA bot.

6. Preparing an Offer Approval for a job application

This is a stage of the job application where offer approvals are prepared for candidates, and it makes a good use case for an RPA bot especially when the job position has mass hiring. Figure 6-10 shows the screenshot where the job offer approval can be initiated.

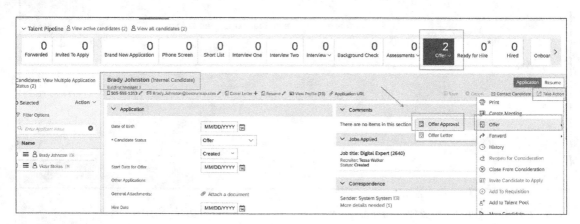

Figure 6-10. *Screenshot Showing Where an Offer Approval for a Candidate Can Be Initiated*

In Figure 6-11, we have advanced two candidates to the offer approval stage. Each candidate requires an offer approval before they can be considered for hiring. Figure 6-10 shows only two candidates in the Offer status. In a real-life situation, especially in case of job requisitions meant for mass hiring, you could have hundreds and sometimes even thousands of candidates for one position. In such cases, an RPA bot can help automate some of the tasks. Figure 6-11 shows the screenshot where the offer approval can be prepared.

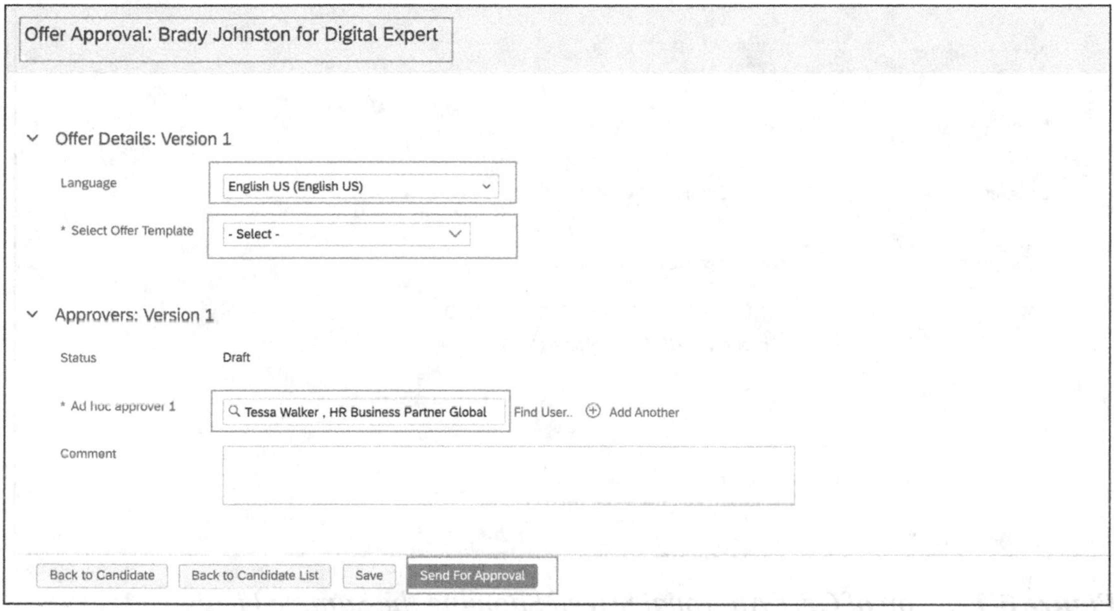

Figure 6-11. *A Sample Offer Approval Screen*

This screenshot is from a demo system. In an actual customer tenant, you could have 10–15 languages and tens of offer templates based on the locale and position. Multiply that combination with hundreds of candidates for a position and you quickly realize that a manual solution will be time consuming. In addition to the selection of the language and the offer template, you may also encounter multiple levels of approvals. While these combinations can quickly result in many variations, you must use some judgment on when an RPA bot will make sense. The higher the job position is in an organization, the more is the manual work in making job offers. Figures 6-12 and 6-13 show the fields often encountered with offer approvals. Of course, the actual screens will vary depending on the customer configuration.

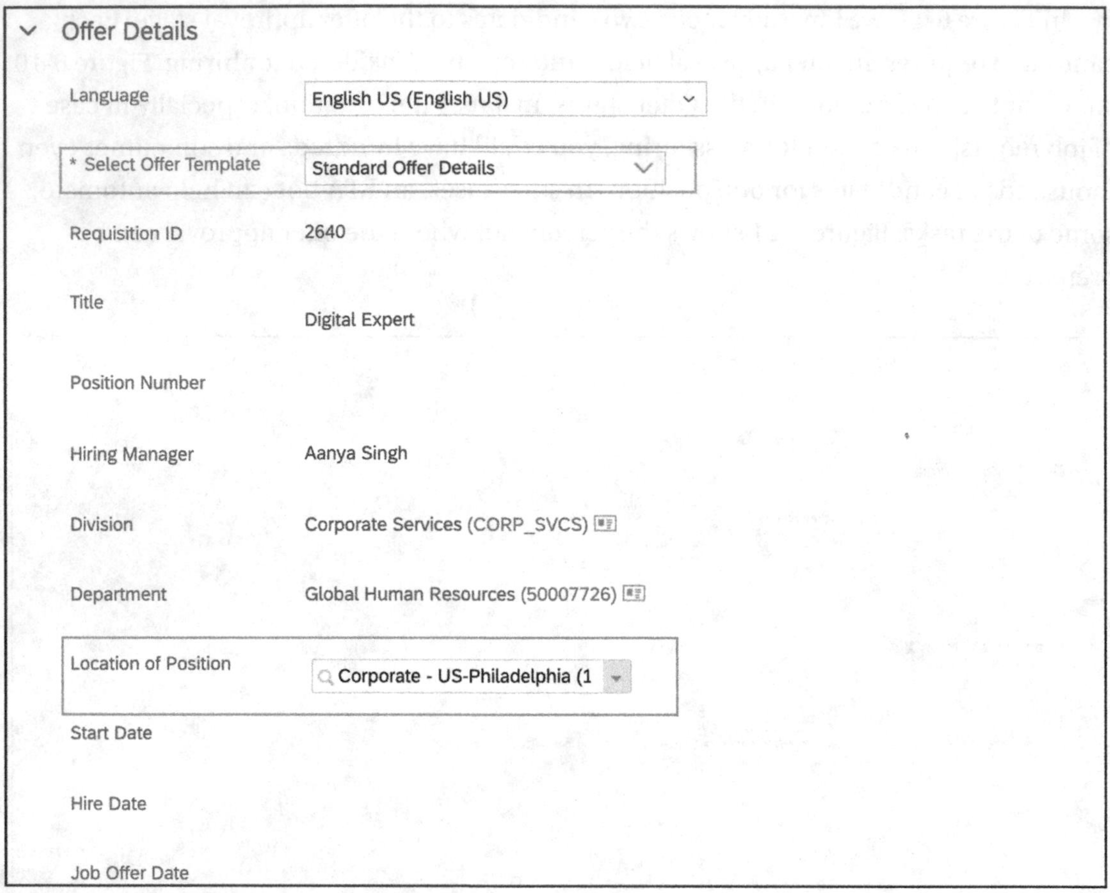

Figure 6-12. *Top of Offer Approval Screen Showing the Sample Fields*

Your RPA bot may have to accommodate the fields shown in Figures 6-12 and 6-13. You must consult with the customer on the use of these fields.

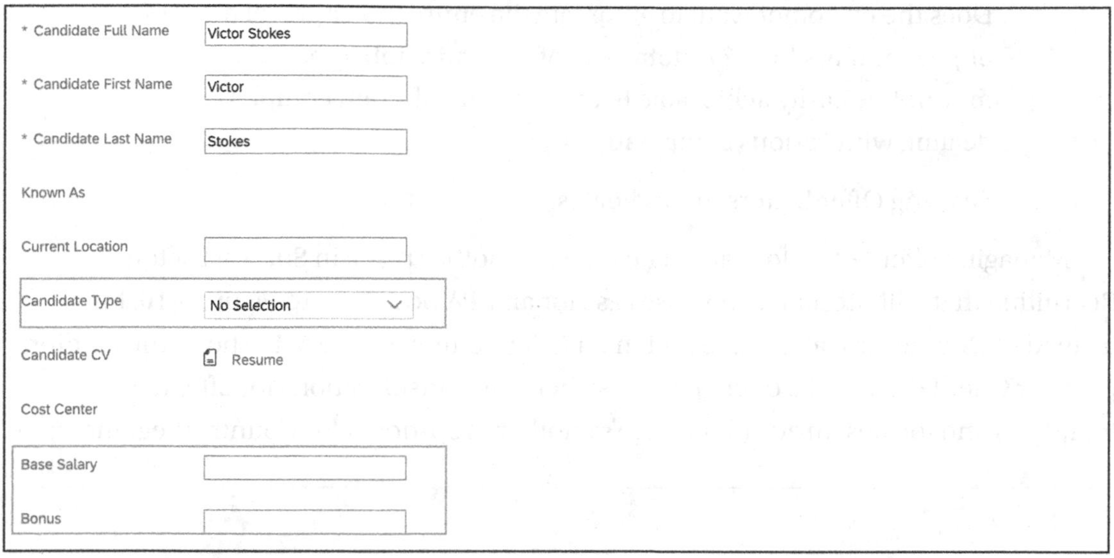

* Candidate Full Name	Victor Stokes
* Candidate First Name	Victor
* Candidate Last Name	Stokes
Known As	
Current Location	
Candidate Type	No Selection ⌄
Candidate CV	🗎 Resume
Cost Center	
Base Salary	
Bonus	

Figure 6-13. *Additional Fields on the Job Offer Approval Template*

Here are some discussion points to have with the customer to make your RPA bot for offer approvals efficient:

- Is the pain point for customers for mass hiring situations or for more complex jobs? The decision will drive the design of your RPA bot.

- Does the customer have multiple offer approval templates for the same position or one template for a mass hiring position? Naturally, the fewer the offer approval templates, the easier will be the design of the RPA bot.

- Does the customer have multiple levels of approval workflow for a mass hire? Very often, mass hires have simple approval workflows. Depending on the customer, you may be able to automate the approval chain too. Keep in mind that you may be able to leverage the Proxy Now functionality offered by SuccessFactors to address.

- Does the customer have multiple versions of the job approval template based on the locale? If that's the situation, you may end up with multiple versions of the RPA bot per locale and job requisition combination.

   –   Does the customer want to automate the entire job approval process
       or just for mass hires? Automation of the entire job approval process
       may not be easily achievable because of the inherent complexity in
       dealing with various combinations.

7.   Sending Offer Letters to candidates

Managing offer letters for candidates can be another place in SuccessFactors
Recruiting that will present a good use case for an RPA bot. Refer to Figure 6-10, but
instead of Offer Approval, use the option for Offer Letters. Figure 6-14 shows the landing
page for Offer Letters. Like offer approvals, there are considerations for offer letter
templates and locales. In addition, there's another drop-down for Country/Regions.

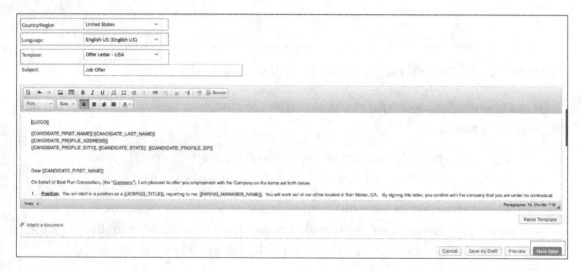

Figure 6-14. *Sample Offer Letter Showing Some Preselected Values*

Figure 6-15 shows the top of the screen for offer letter generation, while Figure 6-16
shows the bottom of the screen. The body of the offer letter is pre-populated based on
the offer template chosen.

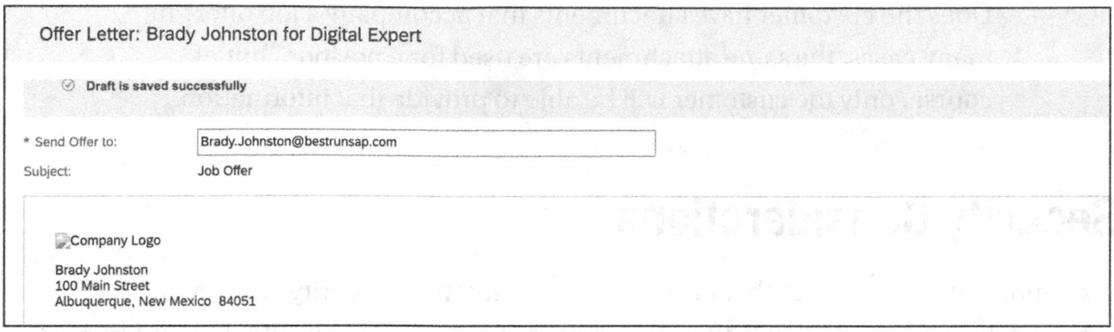

Figure 6-15. *Screenshot Showing the Top of the Screen for Offer Letters*

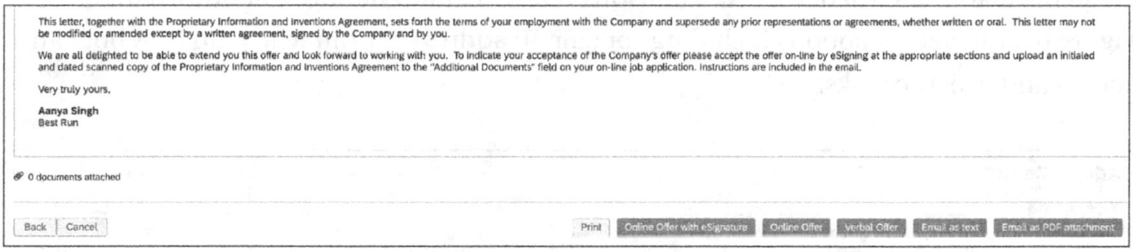

Figure 6-16. *Screenshot Showing the Options Available to Send an Offer Letter to a Candidate*

Here are some discussion points to have with the customer to make your RPA bot for managing offer letters efficient:

- Is the pain point for customers for mass hiring situations or for more complex jobs? The decision will drive the design of your RPA bot.

- Does the customer have multiple job offer templates for the same position or one template for a mass hiring position? Naturally, the fewer the job offer templates, the easier will be the design of the RPA bot.

- Does the customer have multiple versions of the job offer template based on the locale and country/region combination? If that's the situation, you may end up with multiple versions of the RPA bot per locale and job requisition combination.

- Does the customer want to automate the entire job offer process or just for mass hires? Automation of the entire job offer process may not be easily achievable because of the inherent complexity in dealing with various combinations.

- Does the customer have attachments that accompany a job offer? In many cases, the same attachments are used for a position, but of course, only the customer will be able to provide that information.

Security Considerations

It's important to remember that in any customer situation, security is always a consideration. Customers will often create a separate user for your RPA bot and impose restrictions on which IP addresses will be allowed for the RPA bot. Once the user is created, customers will often use "Password & Login Policy Settings" as shown in Figure 6-17 to set exceptions including specific IP addresses from which an RPA bot can access and execute tasks.

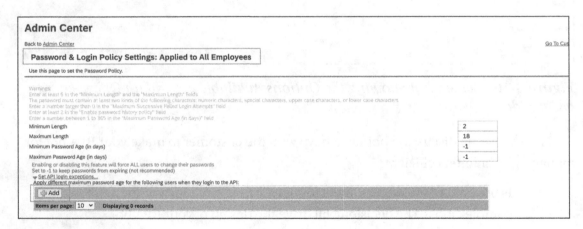

Figure 6-17. *Accessing the Password & Login Policy Settings*

Clicking the "+Add" button shows the screenshot as in Figure 6-18. Customers have the option to add the RPA bot user specifically created and add the IP addresses from where the requests can be made. An advantage of this method is to prevent unauthorized access to the customer's system. You should bring this up when discussing an RPA solution with a customer. The sooner this is addressed up front, the fewer the questions regarding security.

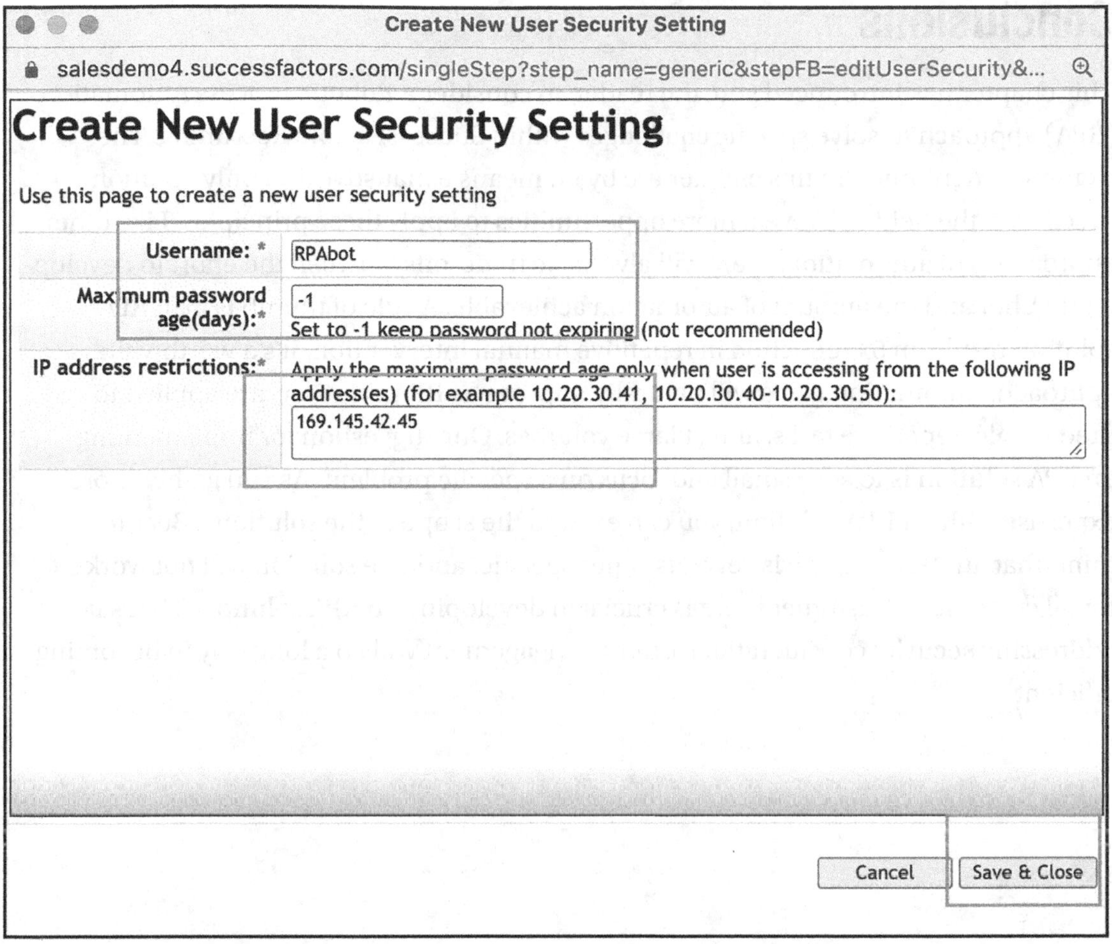

Figure 6-18. *Pop-Up Screenshot to Specify Username and Password*

Assuming the customer has set up a specific user for the RPA bot and has specified IP address restrictions, the customer will probably not share the passwords for established users within SuccessFactors. In such cases, the customers can leverage the "Proxy Now" functionality offered by SuccessFactors and allow the RPA bot to proxy as named users and limit the access to only certain parts of recruiting. An advantage of this approach is that system security is still maintained and yet provides external access to perform specific functions. The audit trail will show that your RPA bot has performed an action on behalf of the named user.

Conclusions

This chapter has introduced you, the reader, to consider a Robotic Process Automation (RPA) approach to solve specific challenges within SuccessFactors Recruiting. The examples mentioned in this chapter are by no means exhaustive and only additional practice in the field will reveal more opportunities to apply these principles. Like other approaches to automation, there will always be trade-offs between the effort to develop an RPA bot and the amount of automation achievable. A rule of thumb is if an RPA solution results in 5x reduction in repetitive manual intervention, it's a worthwhile approach. Automation gains will naturally improve as RPA solutions are applied to predictable, repetitive tasks, and at large volumes. Our suggestion for implementing an RPA solution is to start small and focus on a specific problem. As you gather more expertise with an RPA solution, you can expand the scope of the solutions. Bear in mind that an RPA solution is very customer specific, and one solution will not work for all customers. Customer input is crucial in developing an RPA solution. That said, addressing security considerations prior to engagement will go a long way in becoming efficient.

CHAPTER 7

Automation Using Integration Center

One of the goals of recruiting technology is to free up recruiter time from low-value administrative tasks to better engage with future employees. Recruiting is ripe for automation. In the previous chapter, we looked at how Robotic Process Automation (RPA) tools can be used to reduce the time that recruiters spend on administrative tasks, like adding the same application questions to a requisition again and again. RPA is driven by a predictable set of actions that a user performs, based on rules. Where RPA leverages the user interface of the application to automate processes by replicating user actions, Integration Center uses back-end API calls to affect changes. In this Chapter, we walk you through how to setup such calls step by step.

In Chapter 12, we will talk about how Intelligent Services can use a predefined event to trigger an action, such as the execution of an Integration Center job. Examples of such events include the update of job application information, the approval of an offer, and the update of job application status.

While Intelligent Services events are used to trigger actions in real time, this chapter will focus on how you can update large amounts of recruiting information in batch. Integration Center will be used to do this.

Business rules can also be a good choice for triggering changes to data and will be covered in Chapter 11. An important consideration with updating job application data using business rules is that business rules are not triggered when a candidate applies or changes data. If a mobile-responsive candidate experience is enabled (which it should always be), then business rules are not triggered by the candidate. Business rules will be triggered by HR users such as recruiters when they make changes to application data, such as the application status.

© Anand 'Andy' Athanur, Mark Ingram and Michael A. Wellens 2022
A. A. Athanur et al., *Innovative SAP SuccessFactors Recruiting*, https://doi.org/10.1007/978-1-4842-7425-5_7

Relevant Business Scenarios

Table 7-1 shows examples of possible integration scenarios and the recommended integration method to use.

Table 7-1. *Relevant Business Scenarios*

Requirement	Integration to Use	Reason
If an applicant responds "Yes" to the application question of whether they are a previous employee, then put them in the "Rehire Eligibility Check" status as soon as they apply.	Intelligent Services trigger of Integration Center	This will not have super high volume and supports a real-time solution.
Aspiring baristas apply to a city-wide evergreen requisition and are forwarded to a store-based requisition based on their store location preferences.	Integration Center scheduled job	The volume will be quite high, and this will need to run for all applicants on the evergreen requisition.
Moving an applicant into the "Request Interview Schedule" status sets a reminder flag for employees to talk to their current manager.	Business rule	Because it is a recruiter making the change to the application data, we can automatically trigger a business rule.
If a requisition title contains "Project Manager" and an applicant has the "Project Manager Silver Medalist" tag on their candidate profile, bump them to the "Review Immediately" status upon application.	Intelligent Services trigger of Integration Center	The volume won't be too high for Intelligent Services. Integration Center scheduled job could also be used.

This chapter focuses on the use of Integration Center scheduled jobs. There are several ways of doing scheduled jobs. In our case study, we will use SuccessFactors as both our source and destination system. We will first examine the role of using separate import and export jobs to perform mass updates of information.

Mass Updates of Recruiting Data Using Separate Integration Center Jobs

There are some circumstances where you need to update large amounts of data as a one off. Unlike process automation via a business rule, Intelligent Services event, or scheduled integration jobs, this is run once to solve a specific issue. Examples of use cases are

- Mass update of cost centers on job requisitions

- Creation of historic job applications for migrated employees

- Mass update of locations on job requisitions

- Mass correction of job application fields

Bulk updates using two jobs consist of an export job and an import job, as shown in Figure 7-1.

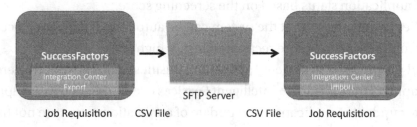

Figure 7-1. *Bulk Update Process Flow*

With the Integration Center, the SAP SuccessFactors solution can be used as both the source and destination system within the same integration job. Why, then, would we create two separate jobs? Using a single job is a good idea for regular batch updates. For one-off updates, the advantage of using one Integration Center job to generate a .CSV file to be consumed by another job is that it gives you the opportunity to check the generated file before importing it.

Using the Integration Center to Automate Recruiting Using SuccessFactors as the Source and Destination System

Why a SuccessFactors to SuccessFactors Integration

Why on earth would you want to integrate an SAP SuccessFactors instance with itself? Typically, it's used to update data in the system using a set of rules. "Wait," you cry. "Isn't that the job of business rules?" Yes it is, normally. In Section 7.4, we'll have a step-by-step case study. Let's look at why we might use SuccessFactors as both the source and destination systems.

Recruiting automation heavily involves moving applications between statuses using some business logic. If I am a recruiter that updates a screening rating with 10/10, then a business rule could be used to automatically move the application to the "Submit to Hiring Manager" status. The business rule is triggered when the recruiter hits save and changes the application status based on the screening score.

The earlier in the process that the system can be automated, the better, because the application volume, and therefore recruiter effort, is higher.

The ideal place to start is at New Application. Business rules are not triggered when candidates apply. There is also no Intelligent Services event triggered by an application. The events for update of application and update of application status are not triggered by new applications. Bummer.

This leaves us with the option of checking for new applications on a regular schedule and updating the application status based on other application data. This could also be achieved by having separate outbound and inbound interfaces, basically one export to say which applications are relevant and another to update the applications. That approach is unnecessary and will create timing issues if application information changes between the export and import jobs running.

Overview of SuccessFactors to SuccessFactors Integrations

Integrations that use the SAP SuccessFactors solution as both the source type and destination type can be best illustrated in Figure 7-2. This is the flow shown on the Review and Run screen of your integration when selecting the SuccessFactors to SuccessFactors integration type.

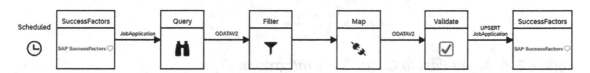

Figure 7-2. *Integration Flow with SAP SuccessFactors Solution as Both Source and Destination Type*

The integration flow performs the following:

- **Query** – Beginning with the starting entity (in our example, this will be JobApplication), include all of the fields that need to be read, either directly from the JobApplication entity or via associated entities such as JobApplication.

- **Filter** – Read just the desired data using filters, such as the status of the job application or the status set of the application (e.g., experienced hires status set vs. intern status set).

- **Map** – Map the source fields to the destination fields. The fields will often be the same, for example, JobApplication-Application ID to Application ID. Some destination fields may be calculated fields or fixed values.

- **Validate** – Before updates to the destination system are made, validation must occur. Examples include a check of whether picklist values are valid, mandatory fields are present, etc.

To create a SuccessFactors to SuccessFactors integration (meaning the same SuccessFactors instance), go to the Integration Center home page, select My Integrations. Then in the top right, click "Create" and then "More Integration Types" as in Figure 7-3.

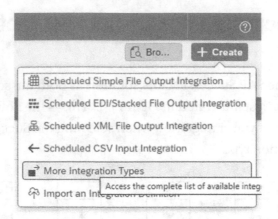

Figure 7-3. *Menu Path to Create More Integration Types*

Select the Trigger Type of Scheduled since we want this to be an ongoing automation. Select SuccessFactors as Destination Type and Source Type. The correct options are shown in Figure 7-4.

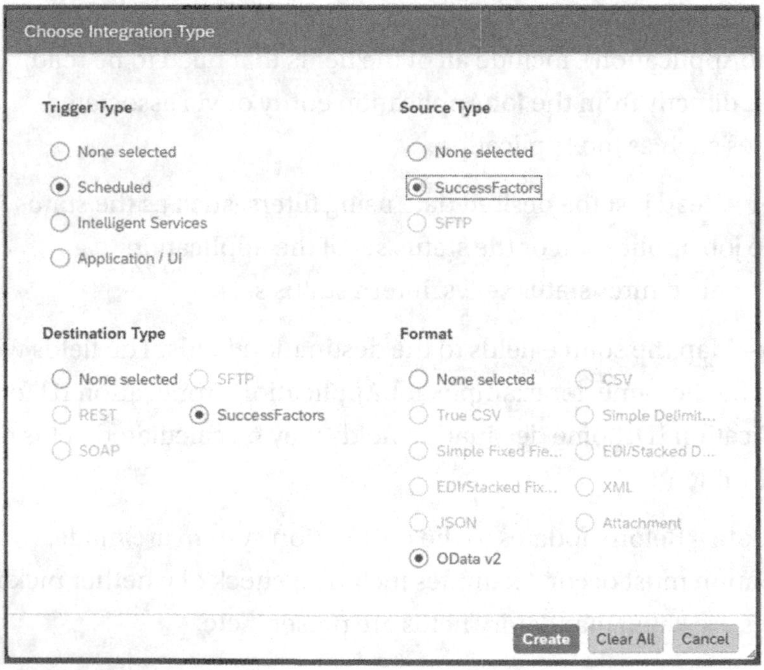

Figure 7-4. *Choose Integration Type*

On the next screen, you select the starting ODATA entities, as we have done in previous chapters. You can search for entities by entity name or field name. When processing applications, JobApplication will likely be our starting entity. Select the entity after it shows in the search results on the left side. All fields and navigations are preselected and cannot be changed. Complete this step by clicking "Select" in the bottom right. The selection of the JobApplication entity is shown in Figure 7-5.

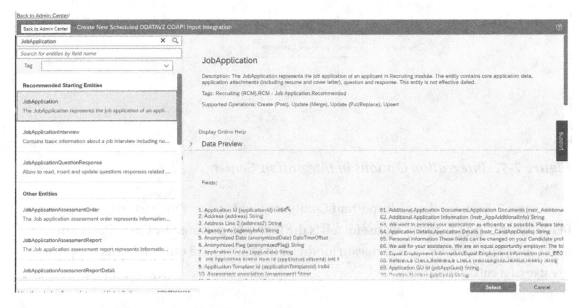

Figure 7-5. *Selection of JobApplication Entity*

Once the starting entity has been chosen, you can see the configuration steps at the top of the screen, shown in Figure 7-6. You will notice that there are no destination settings. This is because you have selected SuccessFactors as both the Destination Type and Source Type.

Figure 7-6. *Configuration Steps in Integration Center*

Figure 7-7 shows the integration Options page. Unlike working with CSV extracts or imports, there are no headers, footers, delimiters, etc. Just specify a name and optionally a useful description.

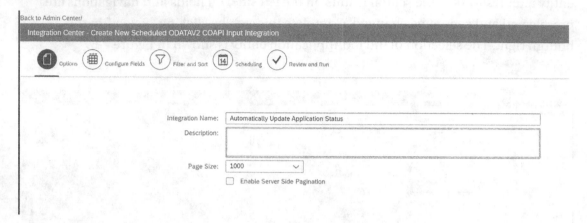

Figure 7-7. *Integration Options in Integration Center*

Figure 7-8 shows the all-important Configure Fields page. Since we are manipulating fields before update, the most useful tab is the first one, denoted by the bullet point icon and called Detailed View. The reason for this is that this is the only view that supports the use of a calculated field.

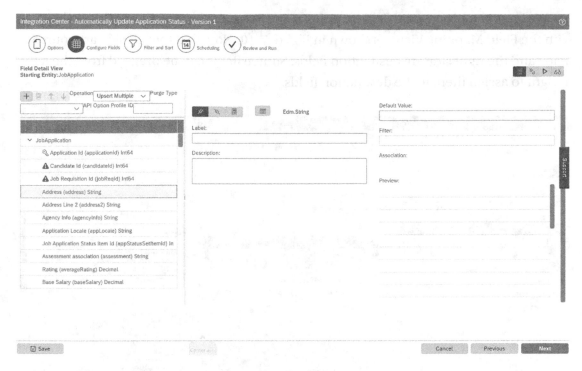

Figure 7-8. *Configure Fields Page in Integration Center*

Note that the destination fields are here shown on the left side.

The field options for populating fields are shown from left to right, as highlighted in Figure 7-9. They are

- **Set as Fixed Value Field** – Set a default value.

- **Set as Associated Field** – Map to another field. This can also be done from the Field Mapping View.

- **Set as Calculated Field** – Perform calculations, If/Then conditions, and more to populate the field.

- **Associate Operand with a Field** – This allows the editing of calculated field, associated fields for mapping, and more.

Figure 7-9. *Icons for Populating Fields in Integration Center*

We have already covered field mapping using Set as Associated Field. This is shown with the Field Mapping View, as shown in Figure 7-10. Note that the left side are source fields and the right side are destination fields. Source fields can be dragged from the left to right to assign them to the destination fields.

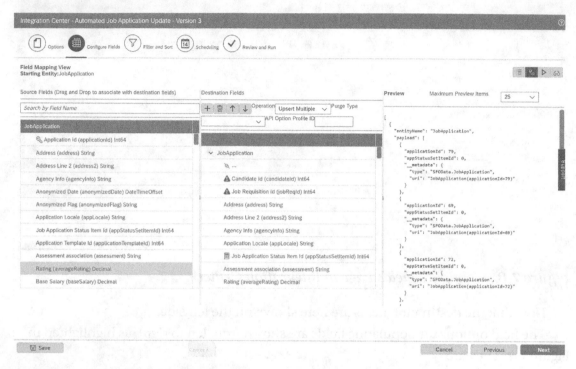

Figure 7-10. *Field Mapping View in Integration Center*

Calculated fields and filters are very important for recruiting automation. Let's take a look at the use of calculated fields. Figure 7-11 shows the different options for calculating field values. Note that there can be multiple rows of statements. Statements can be nested and reside on different levels.

Figure 7-11. *Options for Calculating Field Values*

The options for statements are shown as follows:

Field Value – Set the field value based on a specified field or fields as the operand(s) as well as an operation. Example:

- Set the "Days to Start Day" custom field using Start Date as the operand and Plus Age Calculation as the operator. This could then be used as a dashboard alert for applications of new hires who are nearing their start date.

If/Then – When the If conditions are true, then one or more statements are executed as a block. This may be as simple as setting a value. Example:

- If applications status is "Default" and "Previous Employee" is true, then set status to "Check Rehire Eligibility." As this involves two conditions, the logic looks like this:

  - If Application Status is equal to "Default," then

  - If Previous Employee is equal to true then

  - Set application status to "Check Rehire Eligibility"

If/Then/Else – This is the same as If/Then statements, but a block of different statements can be executed if the condition is determined to be false.

Choose – This sets the field to be one of a series of values, depending on the value of the input field. For those from a programming background, it's like a case statement. It avoids having a lot of nested if/then/else statements. Example:

- For applicants for a consulting job that have specified their willingness to travel on their candidate profile as either 10% or less, 10% to 25%, or 25% to 50%, put them in an "Unwilling to Travel" rejection status.

For Each Entity – This loops through an associated entity such as Application ➤ Candidate ➤ Education and checks values. It is often used for Employee Central, looping through multiple pay components. A recruiting use case could be checking if an applicant has ever worked at a certain competitor and setting an application field based on that.

Case Study: Advancing Candidates Based on Attributes

Our client has a lot of challenges in effectively sourcing certain groups of talent. One such group is highly qualified candidates with project management experience. One of the greatest sources of talent is the existing talent pool of registered candidates and employees within the SAP SuccessFactors Recruiting solution. There are often great runner-up applicants, sometimes referred to as silver medalists, when somebody is hired into a position from a slate of candidates. Recruiters would like visibility into whether an applicant is a silver medalist for another project manager role. They don't want to dig for the information or navigate through several clicks, due to the sheer volume of applications.

Technical Solution

Some options for identifying silver medalist candidates include looking at the Jobs Applied portlet on the application and/or by using a "Project Manager Silver Medalist" tag on the candidate profile. Another great way of grouping and marketing to these candidates is by putting them into a "Project Manager Silver Medalist" talent pool. This talent pool can then be used in targeted email campaigns.

It is not currently possible to filter by a candidates' association to a talent pool using Integration Center. For this reason, we are going to use candidate profile tags to identify silver medalist project managers. If we want to also track such candidates in a list, we can create a talent pool with an overnight search that assigns candidates based on the silver medalist tag. We then have the best of both worlds.

We are going to use Integration Center to do the following:

- Filter applications using the following criteria:

  - Requisition title starts with "Project Manager."

  - Application status is "Default" (New Application).

- For those applications, we will then do the following:

  - Loop through the tags on the candidate profile.

  - If any tags say "Project Manager Silver Medalist," then set the new applicant status to be "Short Listed."

Step 0

Ensure that valid test data is available. The application must be in Default status, the requisition must begin with "Project Manager," and the candidate profile must have the tag "Project Manager Silver Medalist." See Figure 7-12.

Figure 7-12. *New Application on Project Manager Requisition*

Step 1

Create a new integration using "More Integration Types."

Step 2

Select Job Application as the starting entity.

Step 3

On the Filter tab, add Requisition Job Title as the filter field. This is done by clicking jobRequisition ➤ jobReqLocale ➤ Job Title, per Figure 7-13.

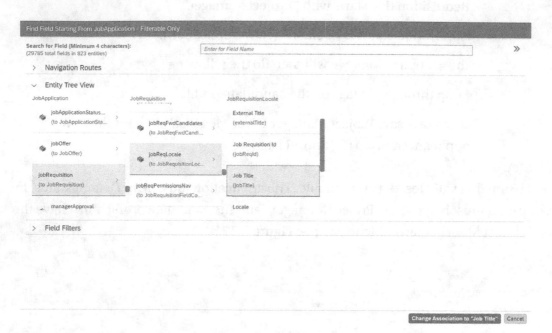

Figure 7-13. *Defining Filter Field*

Step 4

Set the Job Title to start with "Project Manager" and add another filter row using the + button as per Figure 7-14.

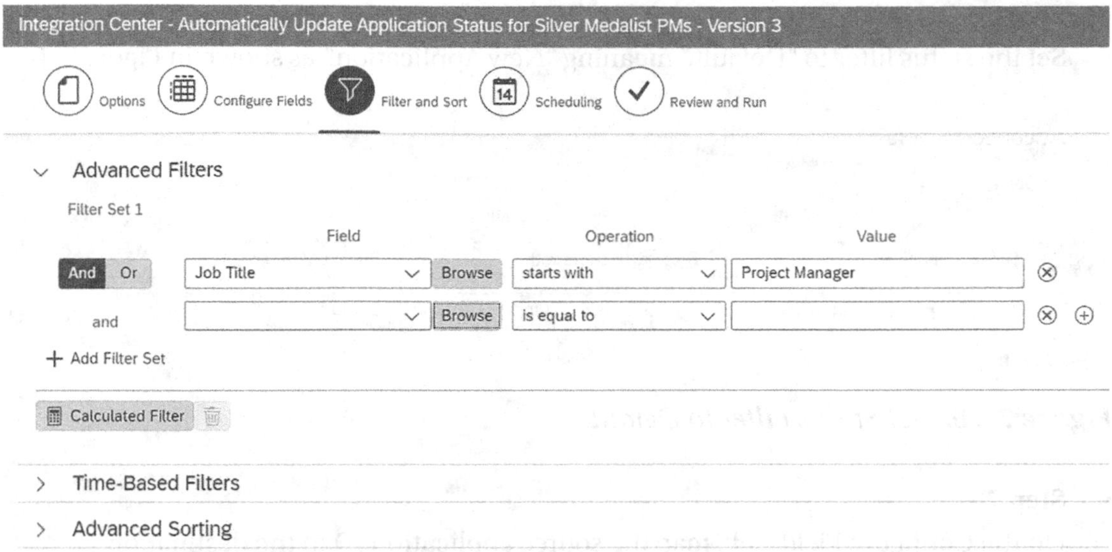

Figure 7-14. *Set Advanced Filters*

Step 5

For the second filter, choose jobAppStatus ➤ Application Status Name as per
Figure 7-15.

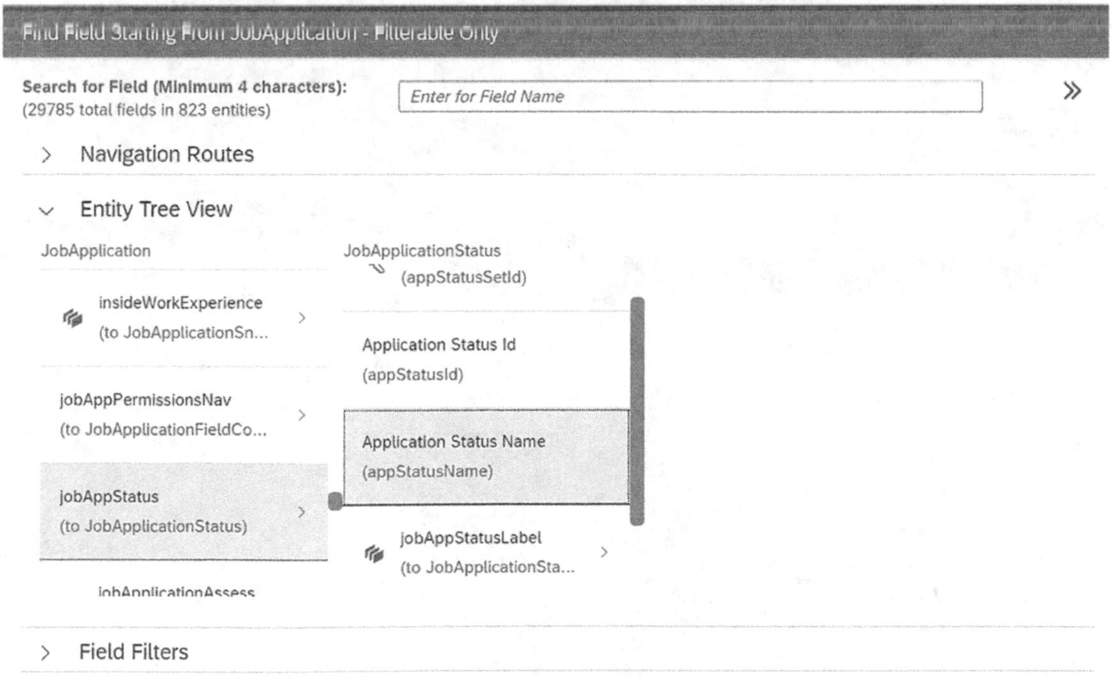

Figure 7-15. *Choose Second Filter of Application Status Name*

Step 6

Set the status filter to "Default," meaning "New Application" as shown in Figure 7-16.

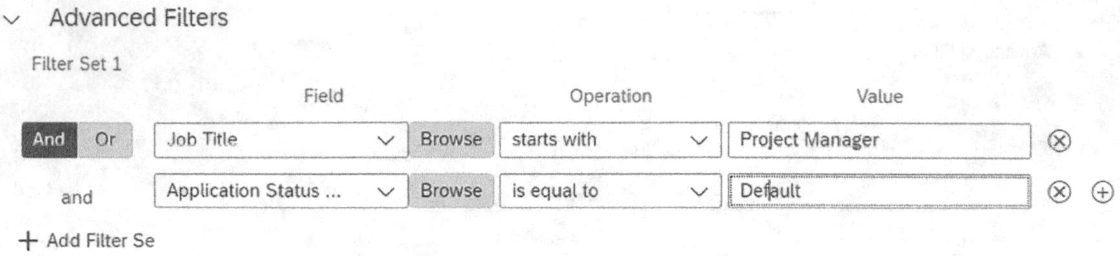

Figure 7-16. *Set Status Filter to Default*

Step 7

On the Configure Fields tab, map the source application ID to the destination application ID.

Step 8

Ensure that you are on the detail view, and click "Job Application Status Item Id" and then click the calculated field icon to the right, as shown in Figure 7-17.

Figure 7-17. *Selecting the Calculated Field Icon*

Step 9

Give the calculated field a label, as shown in Figure 7-18. For the first statement, choose the "For Each Entity" option as we will be looping through the tags associated with the candidate profile tied to the application.

Figure 7-18. Define Field Label

Step 10

Choose the entity to be looped through by clicking <Specify an Entity> and then navigating to Candidate ➤ tags, as shown in Figure 7-19.

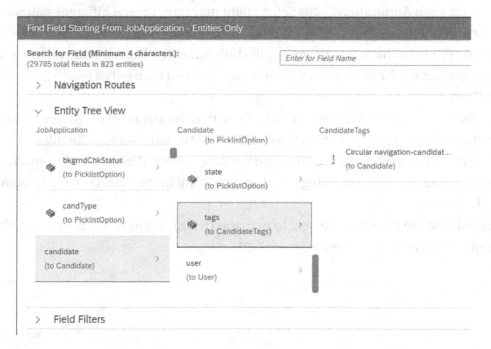

Figure 7-19. Candidate Tags

Step 11

We then check if the tag is "Project Manager Silver Medalist" by using the If/Then statement within the For Each Entity loop. See Figure 7-20.

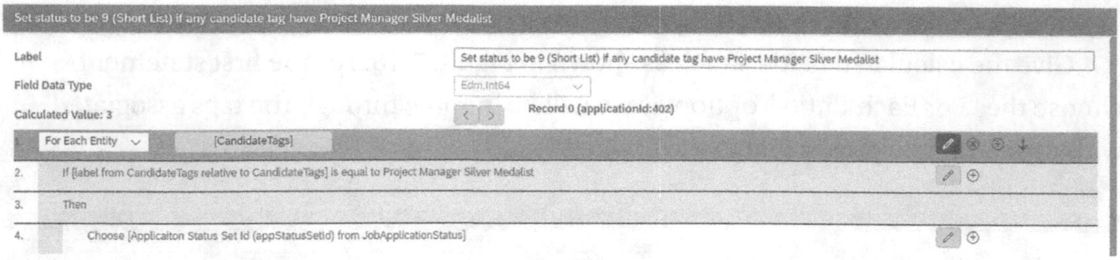

Figure 7-20. Defining the If/Then Statement

Step 12

We want to ensure that applications tagged as silver medalists jump the line and are moved directly to the Short List status. To do that, we update the application status using the application status ID. Application statuses belong in a Status Set, which represent a unique recruiting process. The status IDs for both Default and Short List status will be different for each Application Status Set. A client may have several different status sets to support different processes.

We are going to check the current application status ID. The number will let us know which status set it resides in, as they are unique. That will then tell us which "Short List" status ID to use.

For example, a Default status ID of 63 tells us that the status set is "Simple Status Set," which has the code of 21. The "Short List" status in that status set is 68. In this example, we need to set the new status to 68 to move the candidate to the Short List status. If we try to set the status ID to something that doesn't belong in this status set, the integration job will error out.

We will use the Choose statement to map the Default status IDs to the correct Short List status IDs. This can be seen in Figure 7-21.

Figure 7-21. Mapping Status IDs to Status Set

Step 13

If the status ID isn't recognized in our Choose statement, then it remains zero. After we have exited the For Each Entity loop, we will check if the status ID is still zero. If it is, we will set the new status to be the same as the old one. This means that the application will remain in the Default status. This is shown in Figure 7-22.

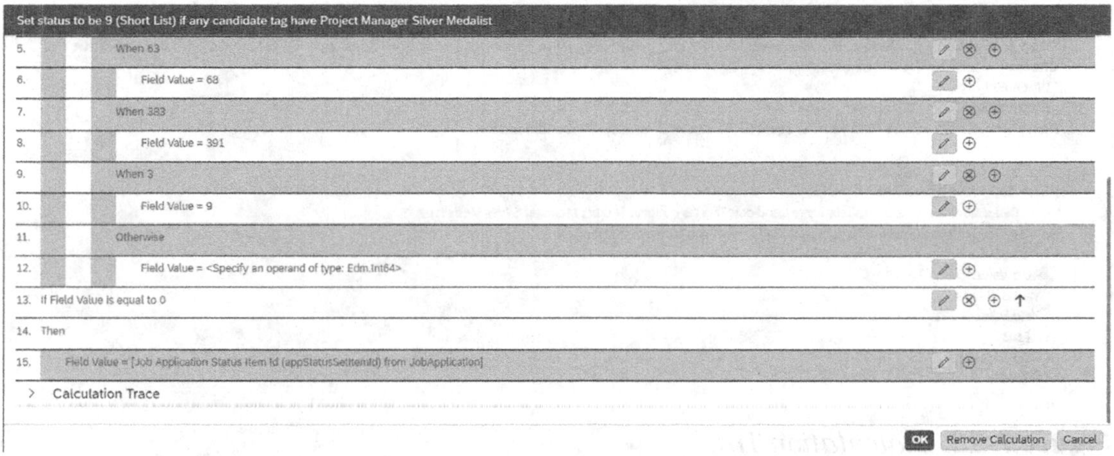

Set status to be 9 (Short List) if any candidate tag have Project Manager Silver Medalist		
5.	When 63	✎ ⊗ ⊕
6.	Field Value = 68	✎ ⊕
7.	When 383	✎ ⊗ ⊕
8.	Field Value = 391	✎ ⊕
9.	When 3	✎ ⊗ ⊕
10.	Field Value = 9	✎ ⊕
11.	Otherwise	
12.	Field Value = <Specify an operand of type: Edm.Int64>	✎ ⊕
13.	If Field Value is equal to 0	✎ ⊗ ⊕ ↑
14.	Then	
15.	Field Value = [Job Application Status Item Id (appStatusSetItemId) from JobApplication]	✎ ⊕
>	Calculation Trace	

OK Remove Calculation Cancel

Figure 7-22. *If/Then Logic*

Those are our steps for setting up the calculated field, other than testing it. The logic of the calculated field can be checked by using the preview records (in this case Applications) and checking how each one is processed. This is done using the Calculation Trace.

Use of the Calculation Trace

The Calculation Trace is very handy for troubleshooting. We will use it to verify that an application that we know should move to "Short List" has generated the correct status ID. For each entity, the logic is shown in the following, along with the evaluation of any conditions as being true or false. In Figure 7-23, we can see that there is a relevant silver medalist tag and that because the status ID for Default is 63, we then set the new status ID to be 68 (Short List).

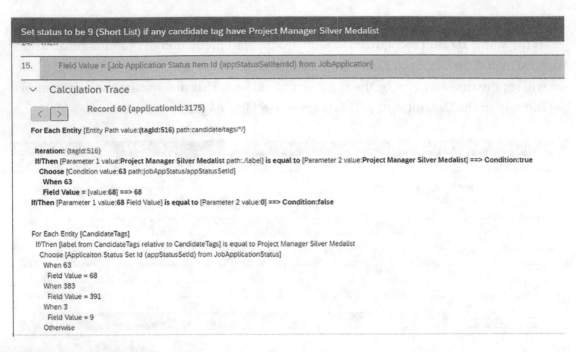

Figure 7-23. *Calculation Trace*

Note   If there aren't enough preview records to test all of your logic, the number of records can be expanded by backing out, going to Configure Fields ➤ Field Mapping View and changing the Maximum Preview items, as shown in Figure 7-24.

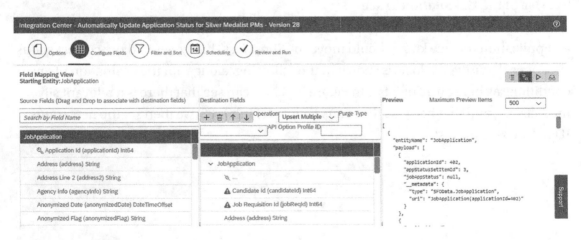

Figure 7-24. *Adjusting the Maximum Preview Items*

Step 14

After we have saved the integration, we will select the Review and Run tab and click Run Now, as shown in Figure 7-25. We can follow the progress by clicking the refresh icon next to the Last Run Time. If there are any errors, we can troubleshoot them.

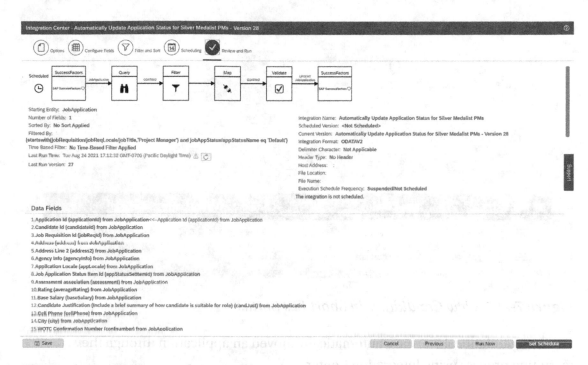

Figure 7-25. *Review and Run Tab in Integration Center*

Step 15

We can now go back to our test application and ensure that it has automatically been moved to the "Short List" status, as shown in Figure 7-26. Once we have successfully completed further testing, we should move the integration to the production environment and schedule it.

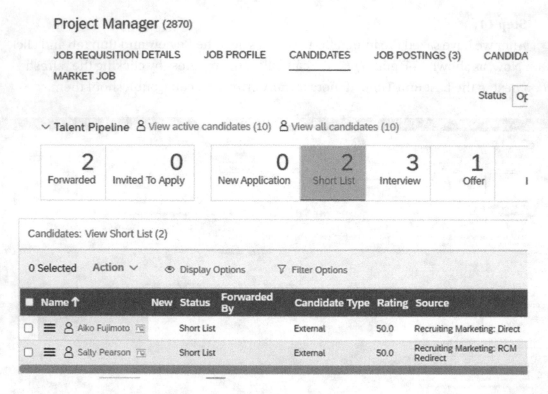

Figure 7-26. *View Candidates in Short List Status*

Congratulations! You have automatically moved an application through the recruiting process using Integration Center.

Conclusion

You have now learned the following:

- How to determine the right tool for applicant automation. Intelligent Services events are great for real-time movement. Integration Center scheduled jobs should be used for high-volume processing.

- Using SAP SuccessFactors as both the source and destination system simplifies automated updates and foregoes the need for SFTP.

- How to perform complex filtering such as the job title based on the job requisition locale.

- Calculated fields are an incredibly powerful tool for using logic to determine how fields are updated.

- The differences between If/Then statements in calculated fields, For Each Entity loops, and the Choose statement.

- How to use the Calculation Trace to know exactly how the logic is being processed for your preview records.

CHAPTER 8

Assessment Integration Framework

In the last few chapters, we've covered how to automate your SAP SuccessFactors system during candidate evaluation. In this chapter, we focus on how to integrate with assessment vendors. The industry has a multitude of assessment vendors to serve the needs of customers geographically, for specific job positions and industries with proprietary implementations. SAP SuccessFactors Recruiting has defined a framework that abstracts the implementation details of each vendor and provides a standard way to integrate. Assessment vendors can help companies greatly in choosing the most qualified candidates in a timely manner using standardized tests tailor-made for specific job positions and incorporating assessments within the applicant workflow. Adding an assessment vendor frees up a recruiter's time to focus on the application process itself by delegating the task to different assessment vendors.

Traditionally, SAP SuccessFactors Recruiting had two assessment vendors, SHL and PeopleAnswers, implemented using Boomi or SAP Cloud Platform Integration (CPI) middleware. The SAP SuccessFactors Recruiting product management team released an assessment framework in mid-2017 for setting up a direct point-to-point assessment integration using the legacy web services-based SFAPI framework. In late 2019, this framework was upgraded to the latest ODATA framework.

Step 1: Provisioning – Check Provisioning Settings

Note As you read through the document, please be advised that the back-end Provisioning tool is only accessible by partners and SAP support.

© Anand 'Andy' Athanur, Mark Ingram and Michael A. Wellens 2022
A. A. Athanur et al., *Innovative SAP SuccessFactors Recruiting*, https://doi.org/10.1007/978-1-4842-7425-5_8

Let's assume that you're a partner implementing a tax screening assessment as part of WOTC (Work Opportunity Tax Credit). By following the steps described in this chapter, you will be able to set the customer's instance to accept WOTC qualification. You will be performing several steps, some of which are done in Provisioning and some in Admin Center. We have called out where each step will be set up.

Navigate to Provisioning ➤ Company Settings. Under Recruiting settings, ensure that Assessment Integration is enabled as shown in Figure 8-1. This step is probably not required, but you should check it, nevertheless.

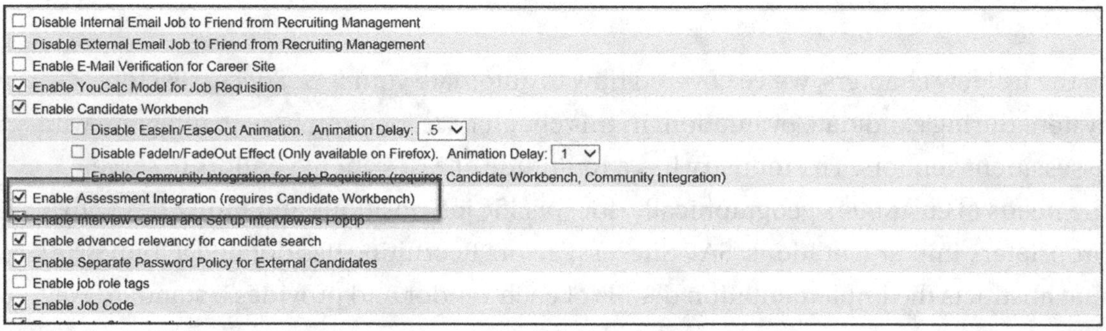

Figure 8-1. *Check Provisioning Settings*

Step 2: Provisioning – Enable Assessment Integration Feature

You have to enable the feature permission for assessments on the job requisition template and for each applicable status. In the following example, assessment integration is enabled for the "Default" status. Ensure that you upload the modified job requisition template(s) in Provisioning.

```
<feature-permission type="assessmentIntegration">
    <description><![CDATA[The following roles can launch interview
    assessment during statuses with an Assessment category]]>
    </description>
    <role-name><![CDATA[S]]></role-name>
    <role-name><![CDATA[O]]></role-name>
    <role-name><![CDATA[R]]></role-name>
    <role-name><![CDATA[G]]></role-name>
    <role-name><![CDATA[V]]></role-name>
```

```
    <role-name><![CDATA[Q]]></role-name>
    <status><![CDATA[Default]]></status>
  </feature-permission>
```

Step 3: Provisioning – Optional Configuration of hardstopStatus

The hardstopStatus is a standard field provided by recruiting that is defined as a picklist called jobReqStatus but treated as a status in the applicant workflow. This feature prevents a job application from proceeding further in the applicant pipeline unless the assessment is completed. The actual status where this hardstopStatus should be enforced can be set in the user interface when you create a job requisition. See Figure 8-15 for an example.

Since it behaves like a standard field, its permissions have to be set in the job requisition like any other field. The field definition is shown as follows:

```
<field-definition id="hardstopStatus" type="picklist" required="false"
custom="false">
  <field-label><![CDATA[Hard Stop Status]]></field-label>
  <field-label lang-"en_US"><![CDATA[Hard Stop Status]]></field-label>
  <field-description><![CDATA[Hard Stop Status]]></field-description>
  <field-description lang="en_US"><![CDATA[Hard Stop Status]]></field-
  description>
    <picklist-id>jobReqStatus</picklist-id>
  </field-definition>
```

Remember to upload your job requisition once you've made the changes.

Step 4: Admin Center – Permissions for Assessment Integration

Two sets of permissions are required for assessments, Administrative Permissions to enable the assessment association and the API user permissions. In most customer environments, these are two separate roles. Table 8-1 summarizes the permissions required for an assessment integration. Some customers may have a single role or separate roles depending on their particular situation.

Table 8-1. *Summary of Permissions Required for Assessment Integration*

System Role	Permission Category	List of Permissions
System Administrator	Manage Recruiting	– Manage Assessment Vendors
API User	General User Permissions	– User Login – SFAPI User Login
API User	Recruiting Permissions	– ODATA API Application Create – ODATA API Application Export – ODATA API Application Audit Export – ODATA API Application Update
API User	Manage Integration Tools	– Access to Event Notification Subscription – Access to Event Notification Audit Log – Access to Outbound Trust Manager

Step 5: Admin Center – Importing/Exporting the Assessment Vendor File

In Admin Center, navigate to Manage Assessment Vendors.

Admin Center

Back to Admin Center

Import/Export Assessment Vendors

○ Import

Choose File: [Choose File] No file chosen

◉ Export

[Submit]

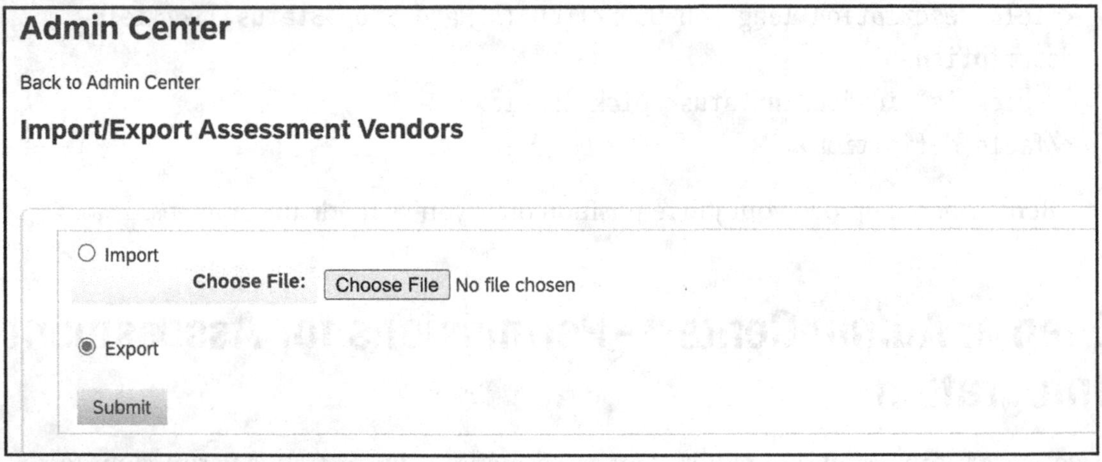

Figure 8-2. *Exporting Assessment Vendor File*

Select Export option as shown in Figure 8-2 and click submit to download the CSV file. You may choose to make an existing vendor active or use your own. Import the file again. Screenshot of a sample vendor is shown in Figure 8-3.

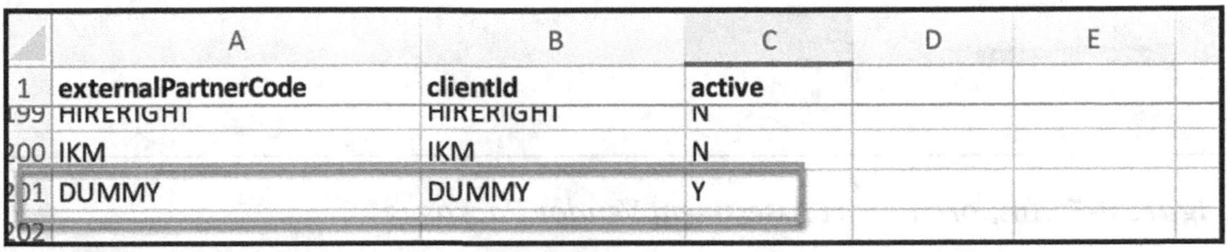

	A	B	C	D	E
1	**externalPartnerCode**	**clientId**	**active**		
99	HIRERIGHT	HIRERIGHT	N		
200	IKM	IKM	N		
01	DUMMY	DUMMY	Y		
02					

Figure 8-3. *Importing Assessment Vendor File*

Step 6: Provisioning – Upload Vendor Assessment Packages

Scroll down to locate the following section.

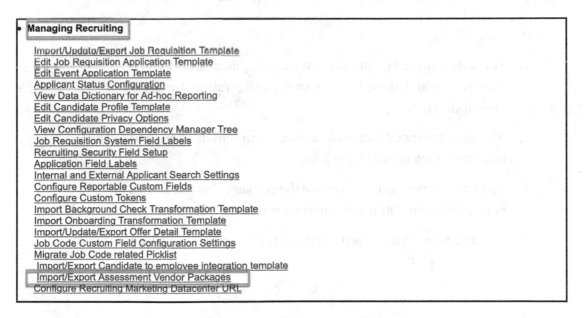

Figure 8-4. *Provisioning – Import/Export Assessment Vendor Packages*

Export the existing vendor package template as shown in Figure 8-4. Complete the assessment vendor package file as shown in Figure 8-5 and import the file.

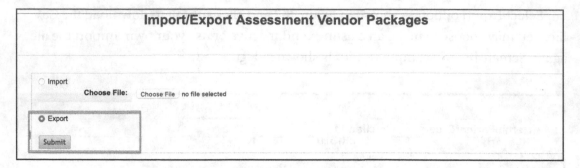

Figure 8-5. *Import/Export Assessment Vendor Packages*

An example of a completed CSV file is shown in Figure 8-6.

A5		fx						
	A	B	C	D	E	K	O	T
1	vendorId	packageCode	reportPackageCode	comparisonGro	shortName	en_US	de_DE	
2	DUMMY	12345	DUMMY_12345_RC		12345_RC	Reading and Comprehension	Lese- und Verständniskeit	
3	DUMMY	98765	DUMMY_98765_JAVA		98765_JAVA	Java Programming - Basic	Java-Programmierung - Stufe 1	
4								

Figure 8-6. *Sample Assessment Vendor Package File*

Notes:

1. The only required columns are vendorId, packageCode,
    shortName, and, depending on your configuration, the localized
    descriptions.

2. Packages are specific to each customer implementation, and once
    imported, they cannot be deleted.

3. As a vendor, you will only receive the packageCode as shown in
    Figure 8-6. shortName column is shown in the user interface.

4. Any additions to existing packages require provisioning access.

Step 7: Admin Center – Setting Up Event Notification Subscription

Navigate to Admin Center ➤ Tools ➤ Event Notification Subscription.

The first step in "Event Notification Subscription" is to define a subscriber. You can add a new one by clicking "Edit Subscriber." The only significant fields are the Subscriber Id, and you can set them to the same value.

Figure 8-7 is an example of a completed subscriber.

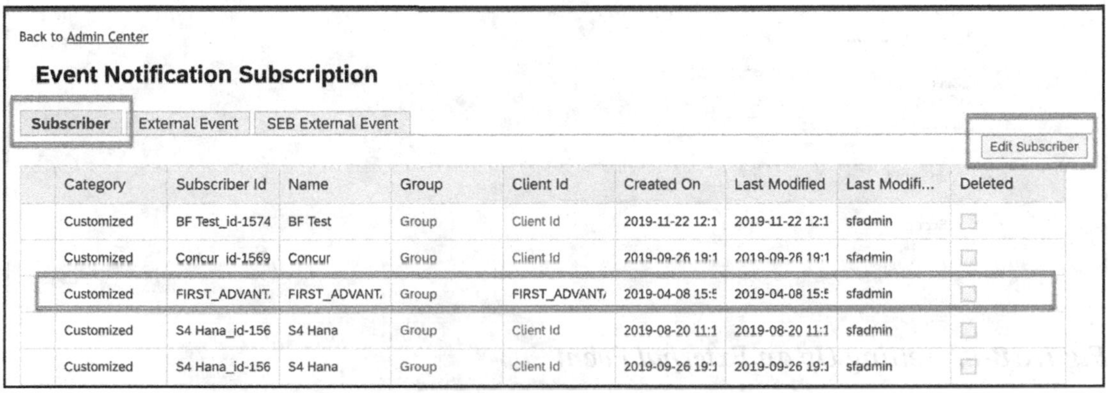

Back to Admin Center

Event Notification Subscription

	Category	Subscriber Id	Name	Group	Client Id	Created On	Last Modified	Last Modifi...	Deleted
	Customized	BF Test_id-1574	BF Test	Group	Client Id	2019-11-22 12:1	2019-11-22 12:1	sfadmin	☐
	Customized	Concur id-1569	Concur	Group	Client Id	2019-09-26 19:1	2019-09-26 19:1	sfadmin	☐
	Customized	FIRST_ADVANT.	FIRST_ADVANT.	Group	FIRST_ADVANT/	2019-04-08 15:5	2019-04-08 15:5	sfadmin	☐
	Customized	S4 Hana_id-156	S4 Hana	Group	Client Id	2019-08-20 11:1	2019-08-20 11:1	sfadmin	☐
	Customized	S4 Hana_id-156	S4 Hana	Group	Client Id	2019-09-26 19:1	2019-09-26 19:1	sfadmin	☐

Subscriber · External Event · SEB External Event · Edit Subscriber

***Figure 8-7.** Event Notification Subscription*

Notes:

1. The important fields to configure are the Subscriber Id (this matches the vendorId in step 5), Name, and Client Id.

2. If the vendor you're connecting to has a common endpoint for all customers, it's a good practice to have a unique client ID for each customer.

Next step is to define the External Event.

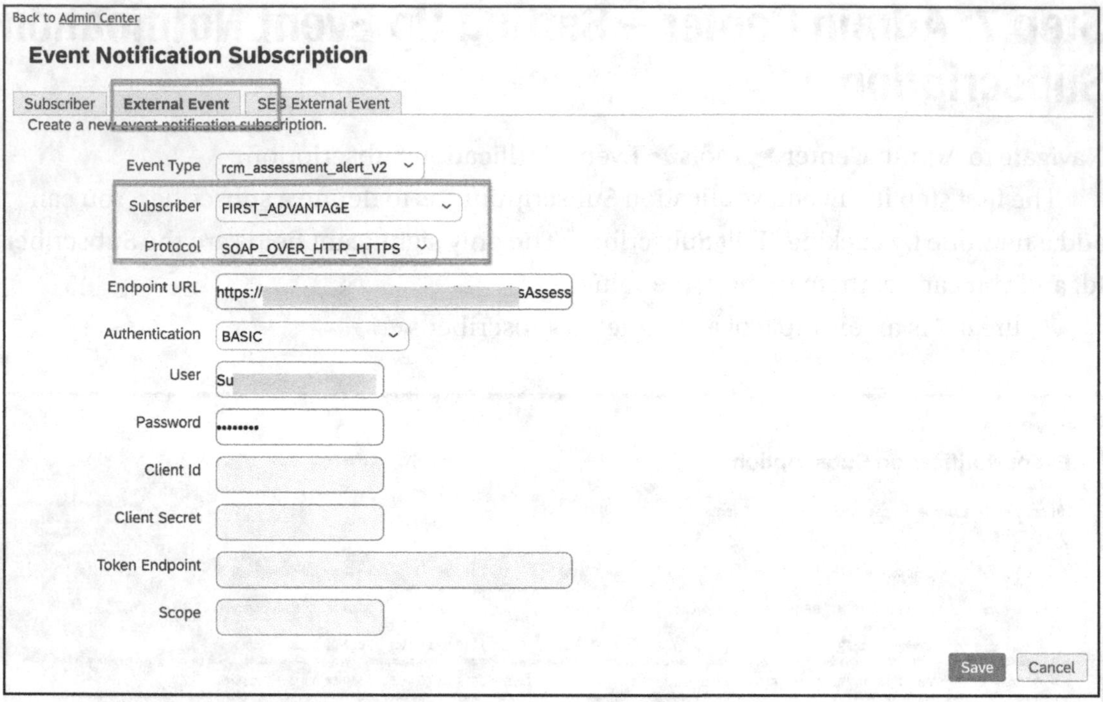

Figure 8-8. *Setting Up an External Event*

Notes:

Event Type – It should be shown as seen in Figure 8-8. The other values in this drop-down are for backward compatibility.

Subscriber – It is the same as the Subscriber Id in the previous screen.

Protocol – It is only SOAP_OVER_HTTP_HTTPS.

Endpoint URL – This is the endpoint of the partner's system that will receive this assessment notification.

Authentication – There are three options, NONE, BASIC, or CLIENT_ CREDENTIALS. Basic Authentication will be deprecated by the end of 2021 and not supported beyond 2022.

The following fields are required if your authentication is BASIC:

> **User** – This is the userId required by the vendor's software to enable integrations from external sources, for example, SAP SuccessFactors Recruiting.

> **Password** – Self-explanatory.

The following fields are required if your authentication employs CLIENT_ CREDENTIALS and is set by your software:

Client Id – A unique identifier on your system to indicate the origin of the assessment request. You should plan on having unique client IDs per customer for productive use.

Client Secret – Similar to the client ID, this is a value known only to you and the customer/implementation partner.

Token Endpoint – This URL is the endpoint where the token is generated based on the client ID and client secret entered.

Scope – It's an optional parameter.

Step 8: Admin Center – Set Up Email Template to Associate with Assessment

Navigate to Admin Center ➤ Manage Recruiting Email Templates.

SAP SuccessFactors Recruiting provides many email templates for you. You can create your own email template or leverage an existing one by modifying it. For this example, we're using a pre-delivered email template called "D6: Assessment details needed" as shown in Figure 8-9.

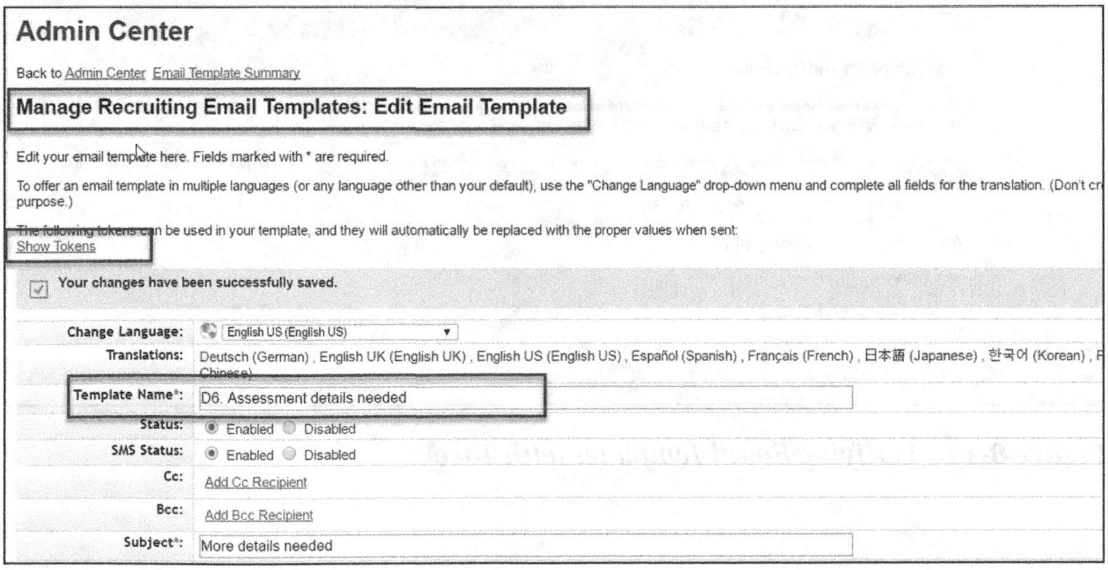

Figure 8-9. *Managing Recruiting Email Templates*

Click Show Tokens and scroll down to ensure that the two tokens are showing up as seen in Figure 8-10.

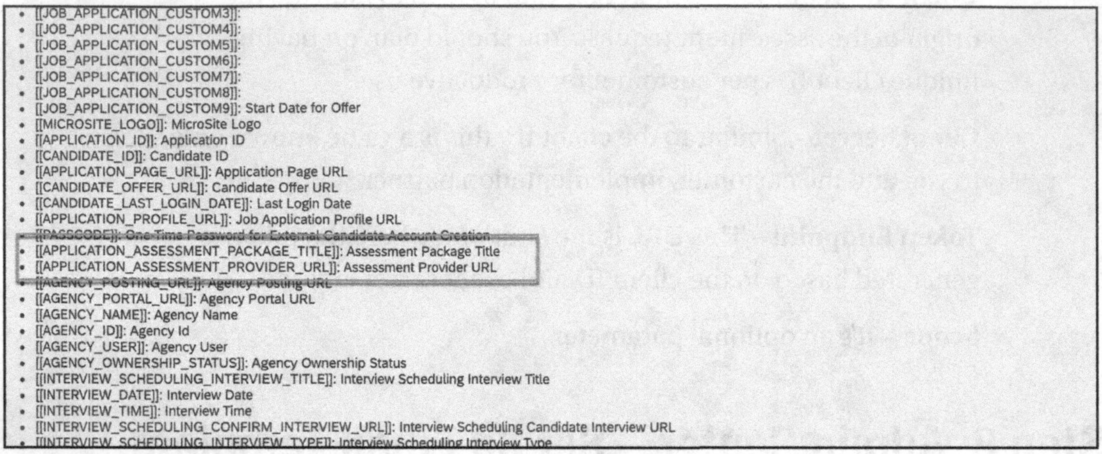

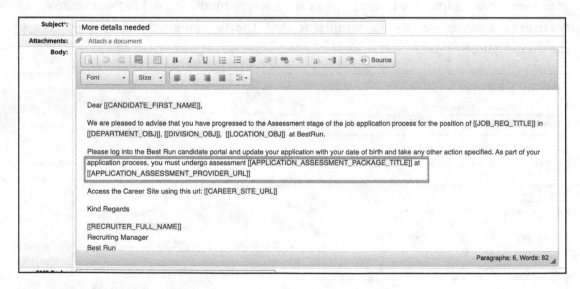

Figure 8-10. *Associating Tokens with Email Templates*

Please ensure that the following tokens are added by clicking Show Tokens as shown in Figure 8-11.

Figure 8-11. *Verifying Email Templates with Tokens*

Note   If your tokens are not highlighted in yellow, it means that the references are incorrect. A quick way to resolve this is to copy and paste the token exactly as shown in Figure 8-10.

Step 9: Recruiting – Creating a Job Requisition with an Assessment and hardstopStatus

This chapter doesn't cover the various considerations while creating a job requisition. Figure 8-12 shows a completed job requisition with the assessment configured in addition to a hardstopStatus.

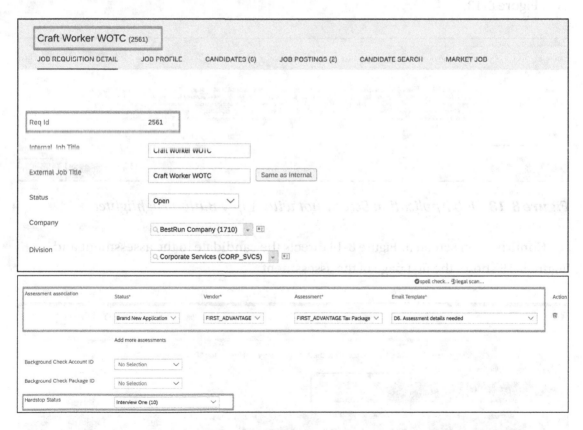

Figure 8-12. *Example of a Completed Job Requisition with Assessment and hardstopStatus*

Step 10: Recruiting – Posting the Job Requisition

Once the job is finalized, you have to post the job requisition. Detailed steps are covered in Chapter 4.

Step 11: Candidate – Undergoing an Assessment

If a job requisition with an associated assessment is posted, the candidate will follow the standard process to register and apply for the job on the customer's career portal. If the job requisition is configured to have an associated assessment in the "Brand New Application" (see Figure 8-12), the candidate is expected to undergo the assessment as soon as the job application is submitted. The candidate is presented with the screenshot as in Figure 8-13.

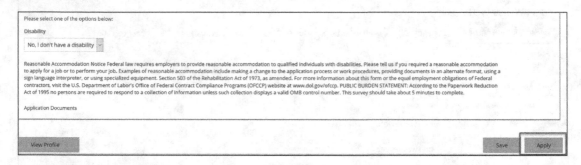

Figure 8-13. *Job Application Screenshot with Apply Button Highlighted*

Confirmation screen in Figure 8-14 directs the candidate to the assessment and Figure 8-15 shows the first page of the assessment.

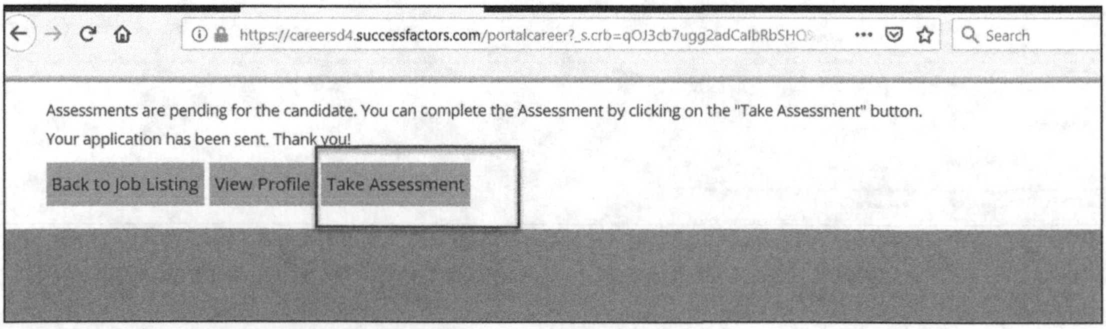

Figure 8-14. *Take Assessment Screenshot*

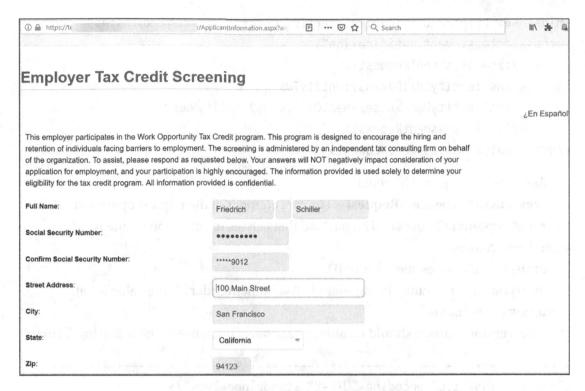

Figure 8-15.   Sample Screenshot Showing First Page of Assessment

Step 12: Candidate – View Email Notifications

If the assessment is configured for a status other than the "Brand New Application," the candidate will receive an email described in step 8. The candidate will then click the APPLICATION_ASSESSMENT_PROVIDER_URL token to undergo the assessment and complete the assessment. The details of the assessment are described in the following.

Step 13: Assessment Order – Request and Response

Upon successful creation of the Assessment Order, SAP SuccessFactors will create a SOAP Request and a synchronous SOAP Response is expected:

```
<?xml version="1.0" encoding="UTF-8" standalone="yes"?>
<ns7:createRcmAssessmentRequest xmlns:ns6="http://schemas.xmlsoap.org/
soap/envelope/" xmlns:ns5="com.successfactors.alert" xmlns:ns7="http://
alert.successfactors.com" xmlns:ns2="http://www.boomi.com/connector/wss"
```

```
xmlns:ns4="http://notification.event.successfactors.com" xmlns:ns3="com.
successfactors.event.notification">
    <ns7:rcmAssessmentRequest>
        <ns7:entityId>182</ns7:entityId>
        <ns7:entityName>AssessmentOrder</ns7:entityName>
    </ns7:rcmAssessmentRequest>
</ns7:createRcmAssessmentRequest>
```

Assessment Request Elements

createRcmAssessmentRequest – The root element of the request operation.

rcmAssessmentRequest – The payload that carries information on the RCM assessment request.

entityId – The assessment order ID.

entityName – The entity being shared, "AssessmentOrder", is the value for all recruiting assessments.

Your vendor software should be able to generate a response in the following format:

```
-------------------------------------------------------------------------
<?xml version="1.0" encoding="UTF-8" standalone="yes"?>
<ns7:createRcmAssessmentRequestResponse xmlns:ns6="http://schemas.xmlsoap.
org/soap/envelope/" xmlns:ns5="com.successfactors.alert" xmlns:ns7="http://
alert.successfactors.com" xmlns:ns2="http://www.boomi.com/connector/wss"
xmlns:ns4="http://notification.event.successfactors.com" xmlns:ns3="com.
successfactors.event.notification">
    <ns7:rcmAssessmentResponse>
<ns7:assessmentUrl>https://test.jobcredits.com/jobcredits/quest/default.
asp?arg=a4c4e89f-d386-4d98-86e2-6cacb042a7b2</ns7:assessmentUrl>
        <ns7:entityId>182</ns7:entityId>
        <ns7:receiptId>a4c4e89f-d386-4d98-86e2-6cacb042a7b2</ns7:receiptId>
        <ns7:status>2</ns7:status>
        <ns7:statusDate>2019-12-19T00:00:00-05:00</ns7:statusDate>
        <ns7:statusDetails>Acknowledged</ns7:statusDetails>
    </ns7:rcmAssessmentResponse>
</ns7:createRcmAssessmentRequestResponse>
```

assessmentUrl – The assessment URL for the candidates. This URL is from a real vendor and is a mandatory field. The assessmentUrl can be queried via the JobApplicationAssessmentOrder entity (step 16).

entityId – The order ID for reference. Mandatory field and it should match the entityId in the request.

receiptId – This is your unique receipt ID for an assessment. You can formulate it any way you want, the primary requirement being that you're able to uniquely identify each assessment request and correlate that with a receipt ID.

status – Mandatory status value. The following enumeration identifies the meaning of each status value:

2 – INITIATED (expected in the initial order acknowledgment)

1 – IN_PROGRESS (expected during status updates)

0 – COMPLETED (expected when an update for assessment completion is received)

99 – INITIATION_FAILED (expected when something goes wrong in processing the assessment request at the vendor side)

statusDate – Date when the transaction is carried out in the standards for XML. Please remember to send the date information in the UTC (with time zone offset).

statusDetails – "Acknowledged" expected for successful order submissions.

Notes:

1. The system expects a synchronous acknowledgment from the vendor within 60 seconds. Otherwise, a timeout will be shown in the audit logs.

2. If for some reason the server is not configured properly on the vendor's end, we will see a generic error indicating "cannot dispatch message." Some errors need to be examined in the server logs.

3. The only XML namespace relevant is the one associated with `http://alert.successfactors.com`, namely, ns7. The actual value of the namespace, "ns7," will depend on the runtime.

Step 14: Recruiting – Viewing Successful Assessments

If your assessment integration is successful, you can check on the progress in two different ways. Figure 8-16 shows the candidate summary page with the assessment status highlighted. If the assessment status is either "Initiated" or "Completed," it's a good sign that your integration is set up correctly.

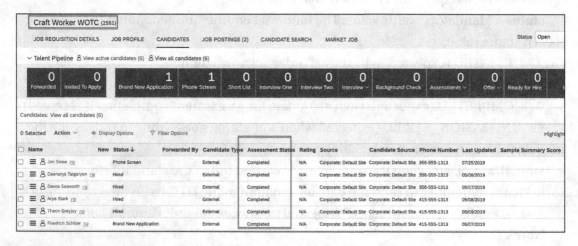

Figure 8-16. *Candidate Summary Page for Assessment Status*

Figure 8-17 shows the details of the results in the Assessments portlet. Click the candidate's name on Figure 8-16 and scroll down to the Assessments portlet of the job application.

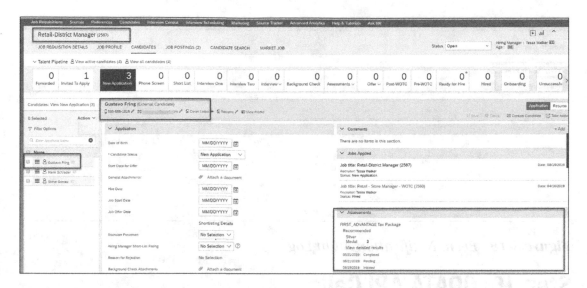

Figure 8-17. *Job Application Details Page with the Assessments Portlet Highlighted.*

In Figure 8-17, the results indicate that the candidate is recommended and has been awarded a "silver medal" with a numerical rating of 3.

Step 15: Admin Center – Troubleshooting Your Integration

Navigate to Admin Center ➤ Event Notification Audit Logs. See Table 8-1 on how to set permissions. If the status says "Error Initiating Order" (Figure 8-19), you should then look at the event audit log. If the following status field indicates DELIVERED, then your integration is successful. If the status indicates FAILED, then you have to click the View hyperlink against each error. The audit log shows the request and response.

Figure 8-18. *Event Notification Audit Log*

Step 16: ODATA API Calls

The endpoint in this example is `https://apisalesdemo4.successfactors.com/odata/v2/JobApplicationAssessmentOrder`.

Querying JobApplicationAssessmentOrder

Query JobApplicationAssessmentOrder to find out details of the assessment order and the relevant job application fields: note these ODATA API calls require Basic Authentication to work. The following samples shown in Figures 8-19 and 8-20 use Postman to perform the API calls.

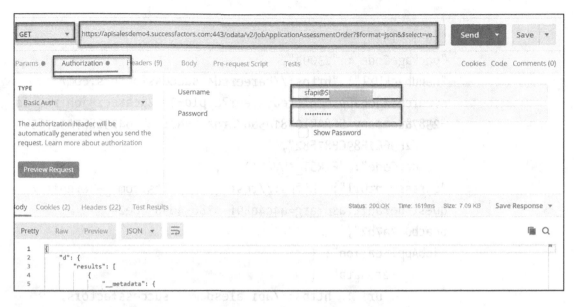

Figure 8-19. *ODATA API GET Call on JobApplicationAssessmentOrder*

```
GET: https://apisalesdemo4.successfactors.com:443/odata/v2/JobA
pplicationAssessmentOrder?$format=json&$select=vendorCode,packa
geCode,id,assessmentUrl,handbackUrl, jobApplication/firstName,
jobApplication/country,jobApplication/lastName, jobApplication/
contactEmail,jobApplication/dateOfBirth,jobApplication/
address,jobApplication/address2,jobApplication/city/jobApplication/
gender,jobApplication/zip,jobApplication/state,jobApplication/
applicationId,jobApplication/status,jobApplication/state/picklistLabels/
label&$filter=id eq '182' and jobApplication/state/picklistLabels/locale eq
'en_US'&$expand=jobApplication,jobApplication/state/picklistLabels
```

Response Payload (only shown partially):

```
{
    "d": {
        "results": [
            {
                "__metadata": {
                    "uri": "https://apisalesdemo4.successfactors.com:443/
                    odata/v2/JobApplicationAssessmentOrder(182L)",
                    "type": "SFOData.JobApplicationAssessmentOrder"
```

```
        },
        "id": "182",
        "packageCode": "13092",
        "handbackUrl": "https://careersd4.successfactors.com/
        sfcareer/jobappcompletedcareer?appId=3022&career_job_req_id
        =2587&company=SFPARTO38145&username=&st=28F6429CCF7A9A7C5DA
        4D45F2F96C1F89CB82F82",
        "vendorCode": "FIRST_ADVANTAGE",
        "assessmentUrl": "https://test.jobcredits.com/jobcredits/
        quest/default.asp?arg=a4c4e89f-d386-4d98-86e2-
        6cacb042a7b2",
        "jobApplication": {
            "__metadata": {
                "uri": "https://apisalesdemo4.successfactors.
                com:443/odata/v2/JobApplication(3022L)",
                "type": "SFOData.JobApplication"
            },
            "applicationId": "3022",
            "zip": "85052",
            "firstName": "Gustavo",
            "country": "United States",
            "lastName": "Fring",
            "address": "300 Los Pollos Hermanos Plaza",
            "contactEmail": "xxx_yyyy@gmail.com",
            "address2": null,
            "city": "Albuquerque",
            "dateOfBirth": null,
            "status": "Open",
            "state": {
                "results": [
                    {
```

```json
"__metadata": {
    "uri": "https://apisalesdemo4.
    successfactors.com:443/odata/v2/
    PicklistOption(11674L)",
    "type": "SFOData.PicklistOption"
},
"id": "11674",
"minValue": "-1",
"externalCode": "state_United_States_New_
Mexico",
"maxValue": "-1",
"optionValue": "-1",
"sortOrder": 49,
"mdfExternalCode": "state_United_States_
New_Mexico",
"status": "ACTIVE",
"parentPicklistOption": {
    "__deferred": {
        "uri": "https://apisalesdemo4.
        successfactors.com:443/odata/
        v2/PicklistOption(11674L)/
        parentPicklistOption"
    }
},
"picklistLabels": {
    "results": [
        {
            "__metadata": {
                "uri": "https://
                apisalesdemo4.
                successfactors.
                com:443/odata/v2/
                PicklistLabel(locale='en_
                US',optionId=11674L)",
```

```
                                    "type": "SFOData.
                                    PicklistLabel"
                                },
                                "optionId": "11674",
                                "locale": "en_US",
                                "id": "12105",
                                "label": "New Mexico",
                                "picklistOption": {
                                    "__deferred": {
                                        "uri": "https://apisale
                                        sdemo4.successfactors.
                                        com:443/odata/v2/
                                        PicklistLabel(locale=
                                        'en_US',optionId=11674L)
                                        /picklistOption"
                                    }
                                }
                            },
```

Notes:

1. ODATA protocol does not permit a filter on the child entity. The response therefore will contain all locales for the PicklistOptions and PicklistLabels. You can filter out the locales you do not require.

2. handbackUrl is a new property introduced in November 2020 release that is useful to redirect the candidate back to the original location before launching the assessment.

Upsert #1: To Indicate That the Assessment Is in Progress

Figure 8-20. *Upsert Sample to Send Results to SuccessFactors*

POST https://apisalesdemo4.successfactors.com:443/odata/v2/
upsert?$format=json

```
{
"__metadata": {
      "type":"SFOData.JobApplicationAssessmentReport",
      "uri":"JobApplicationAssessmentReport"
    },
    "orderId": "182",
    "reportURL": "https://www.google.com",
    "status": 1,
    "statusDate": "/Date(1566422180219)/",
    "statusDetails": "Order is currently being processed."
}
```

Response payload:

```
{
    "d": [
        {
            "key": "JobApplicationAssessmentReport/id=212",
            "status": "OK",
            "editStatus": null,
            "message": "Assessment Report has been inserted successfully",
            "index": 0,
            "httpCode": 200,
            "inlineResults": null
        }
    ]
}
```

Notes:

1. The orderId field is the primary key in the JobApplicationAssessmentOrder shown in the earlier query.

2. The status field can take the following values:

    a. 2 (INITIATED – shown as Initiated in the UI)

    b. 1 (IN_PROGRESS – shown as Pending in the UI)

    c. 0 (COMPLETED – shown as Completed in the UI)

    d. 99 (INITIATION_FAILED – shown as Error ordering assessment in the UI)

3. statusDate is always passed in Unix epoch time. In the earlier example, 1566422180219 means 8/21/2019, 2:16:20 PM Pacific Daylight Time. Current Milliseconds is an excellent resource for this.

The result can be seen in the Assessments portlet on the candidate profile as shown in Figure 8-21

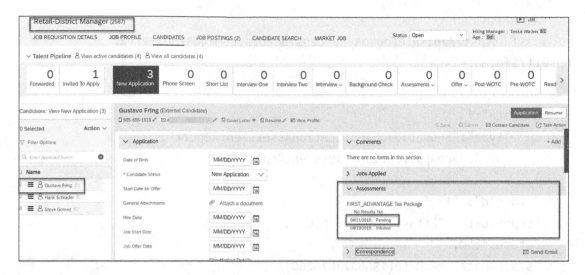

Figure 8-21. *Assessments Portlet Showing a Pending Result*

Upsert #2: To Indicate That the Assessment Is Completed

Figure 8-22 shows a JSON response within Postman, indicating that the assessment result was successfully inserted.

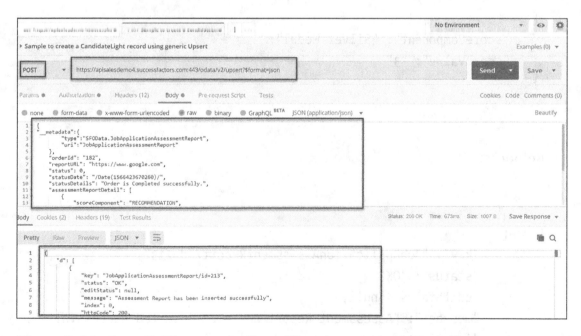

Figure 8-22. *Sample Upsert to Indicate That Assessment Is Complete*

Actual Payload:

POST https://apisalesdemo4.successfactors.com:443/odata/v2/upsert?$format=json

```
{
"__metadata":{
        "type":"SFOData.JobApplicationAssessmentReport",
        "uri":"JobApplicationAssessmentReport"
    },
    "orderId": "182",
    "reportURL": "https://www.google.com",
    "status": 0,
    "statusDate": "/Date(1566423670260)/",
    "statusDetails": "Order is Completed successfully.",
    "assessmentReportDetail": [
    {
        "scoreComponent": "RECOMMENDED",
        "scoreValue": "1"
    },
    {
        "scoreComponent": "Silver Medal",
        "scoreValue": "3"
    }
]
}
```

Response:

```
{
    "d": [
        {
            "key": "JobApplicationAssessmentReport/id=213",
            "status": "OK",
            "editStatus": null,
            "message": "Assessment Report has been inserted successfully",
            "index": 0,
            "httpCode": 200,
```

```
        "inlineResults": null
    }
  ]
}
```

Notes:

1. The orderId field is the primary key in the JobApplicationAssessmentOrder shown in the earlier query.

2. The status field can take the following values:

   a. 2 (INITIATED – shown as Initiated in the UI)

   b. 1 (IN_PROGRESS – shown as Pending in the UI)

   c. 0 (COMPLETED – shown as Completed in the UI)

   d. 99 (INITIATION_FAILED – shown as Error ordering assessment in the UI)

3. statusDate is always passed in Unix epoch time. In the earlier example, 1566423670260 means 8/21/2019, 2:41:10 PM Pacific Daylight Time.

4. The [and] characters indicate that the details are passed as an array to the assessmentReportDetail navigable entity on the JobApplicationAssessmentOrderReport entity. The underlying ODATA entity is JobApplicationAssessmentReportDetail.

5. You may not directly upsert to JobApplicationAssessmentReportDetail entity. You will see an error response as follows:

```
{
    "d": [
        {
            "key": null,
            "status": "ERROR",
            "editStatus": null,
            "message": "Entity JobApplicationAssessmentReportDetail
                        is not editable. Please check the entity
                        setting in Admin Center > OData API Data
                        Dictionary or the entity metadata.",
```

```
            "index": 0,
            "httpCode": 400,
            "inlineResults": null
        }
    ]
}
```

6. scoreComponent and scoreValue on the
    JobApplicationAssessmentReportDetail entity: While this entity
    is an array, it's recommended that the first pair of these values be
    sent as shown earlier.

    The acceptable values are

    a. scoreComponent: RECOMMENDED, scoreValue: 1

    b. scoreComponent: RECOMMENDED_WITH_QUALIFICATION,
        scoreValue: 2

    c. scoreComponent: RECOMMENDED_WITH_RESERVATION,
        scoreValue: 3

    d. scoreComponent: NOT_RECOMMENDED, scoreValue: 4

7. You can send any additional name/value pairs of
    scoreComponent/scoreValue as part of completion.
    There's no recommendation on what values you can send.
    In the previous sample, I've used scoreComponent: "Silver
    Medal," scoreValue = "3".

Step 17: Recruiting – Visual Confirmation of Assessment Report

Figure 8-23 shows the candidate summary page where you can see the status of the assessment for each candidate.

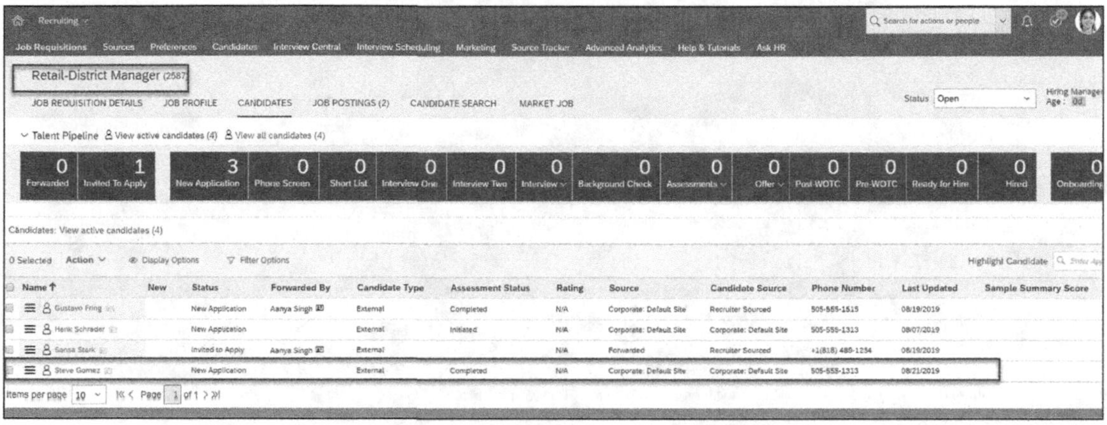

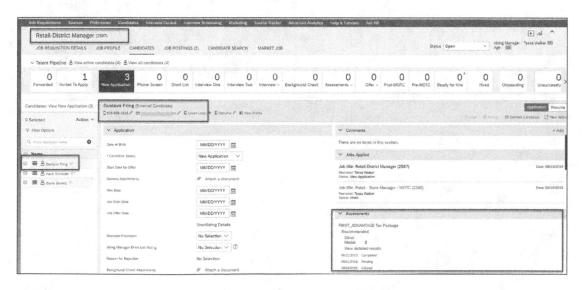

Figure 8-23. *Screenshot to Show Completion of Assessment*

Now click the candidate's name – results are shown in the Assessments portlet as shown in Figure 8-24.

Figure 8-24. *Screenshot to Display Results of Assessment on the Job application Detail*

Conclusion

Congratulations! You have now successfully set up a new assessment vendor in a customer's instance. A good practice is to keep a record of the changes you've made so that you can repeat this process in another customer's instance or to set up a brand-new vendor.

CHAPTER 9

Additional Custom ODATA Integrations

The first part of this chapter will describe the context of when to use SAP SuccessFactors ODATA APIs when working with other solutions within the recruiting industry. The rest of the chapter will be used to explore additional ODATA use cases.

In the previous chapters, we have explored the key entities to be used when working with recruiting integrations. These have included job requisitions, job application, and candidates. We've also learned how to filter these entities and navigate to other entities using both Postman and the Integration Center functionality with SAP SuccessFactors. In this chapter, we'll explore additional ODATA entities that are super useful, as well as their use cases. We'll finish with a case study of matching candidates in job requisitions and the talent pool to job requisitions using third-party AI tools.

Use of Open Standards

This book is focused on the use of SAP SuccessFactors technology to build innovative solutions to recruiting problems, using ODATA APIs. Any kind of integration with SAP SuccessFactors requires the use of ODATA for the exchange of data. Though many vendors make direct ODATA API calls to SAP SuccessFactors solutions, SAP Cloud Platform Integration (CPI) is often used to transform ODATA requests and responses to existing APIs that are offered by a third-party solution.

When integrating with other solutions, open standards are used across many industries and technologies. In the HR space, the HR Open Standards Consortium is used as a common vocabulary between vendors, customers, and systems to simplify integration. Formerly known as the HTR-XML Consortium, the consortium has moved from XML to JSON with its 4.x release. This move to JSON further simplifies integrations and increases flexibility. Find more information at `www.hropenstandards.org`.

© Anand 'Andy' Athanur, Mark Ingram and Michael A. Wellens 2022
A. A. Athanur et al., *Innovative SAP SuccessFactors Recruiting*, https://doi.org/10.1007/978-1-4842-7425-5_9

Open standards mean that we have a common way of representing requests between systems such as background check requests, searches for candidates, and online assessment results. There are also common entities that are shared between different types of requests. Examples include candidate, skill, and position opening. Software vendors will often need to pass additional information that is not covered in the standard. For this reason, open standards are extensible. SAP, like other software providers, has contributed greatly to these standards. One of the authors of this book worked on what was known as the Staffing Exchange Protocol, a standard of the HR-XML Consortium back in the early 2000s. He would recommend working with this great community if open HR standards and collaboration are your thing.

If SAP SuccessFactors is to talk to non-SAP systems using industry standards such as those defined by the HR Open Standards Consortium, then SAP Cloud Platform Integration (CPI) is used to transform the representation of data between ODATA and the standard-based third-party solution APIs .

Prebuilt Interfaces to Third-Party Solutions

When integrating with third-party solution providers, a lot of the work may have been done for you, or in some cases all of it. As we have already said in Chapter 4, there are many recruiting solution providers listed in the SAP Store (see Figure 9-1). The presence of a solution within the store indicates there is very likely a prebuilt integration and documentation. It also indicates that at the time of the addition to the store, the solution was not in conflict with offers planned by SAP SuccessFactors. Note the use of "at the time." Roadmaps change.

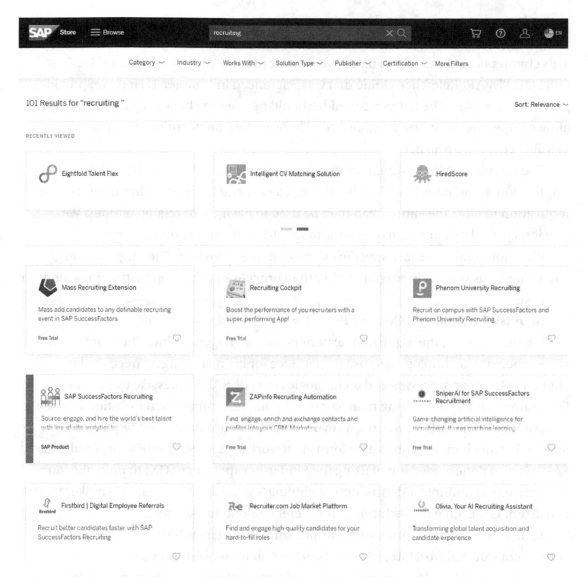

Figure 9-1. *Recruiting Solutions in SAP Store*

Typically, integration guides are provided by the third-party solution provider, in addition to prebuilt integration assets such as Integration Center integration definitions (.icd files) or CPI iFlows. Any custom work that is not prebuilt usually involves additional service costs as well as additional time that needs to be allowed for.

ODATA API Overview

This chapter is not intended to be a replacement to the SAP SuccessFactors HXM Suite ODATA API Reference Guide that was referenced in Chapter 1. Here, we provide practical examples. The Reference Guide should be used to check relevant attributes, allowed operations for listed entities, etc. There are too many entities and use case to possibly cover them all here.

The recruiting API documentation is divided between Job Requisition, Job Application, Candidate, Job Offer, RcmCompetency, and lastly Function Imports. From requisition to offer, the entities can then be used to navigate to related entities. We previously used this to assign questions to requisitions, among other things.

We want to call out some specific use cases in the following. The purpose is to give you some ideas of what is possible rather than provide step-by-step instructions for each use case.

Requisition Approval Custom Front End

For most clients, the requisition experience for managers is fine. There are, however, cases where the technology and user experience for hiring managers needs to be aligned with other processes and technologies outside SAP SuccessFactors solutions. For example, the form a hiring manager uses to initiate or approve requisitions should look similar to the forms used to approve a purchase order or time off request. SAP SuccessFactors is great at providing forms that work for SAP SuccessFactors modules. SAP has also done a good job of using the Fiori user experience (UX) to unify the look and feel across multiple SAP products, including SAP SuccessFactors. This makes for a consistent UX, even if the technology is different behind the scenes.

Sometimes approval forms from various solutions such as SAP Fieldglass for contingent external workforce, SAP Ariba for purchase requisitions, and non-SAP solutions such as Oracle for procurement and ServiceNow for equipment provisioning need to be aligned from a user experience perspective. When this is desired, a third-party technology such as Adobe Forms or ServiceNow can be used. This may also be the case when you want to have a single mobile experience across multiple solutions.

One approach to solving this challenge is to hardcode the forms in the third-party solution. If you take this approach, it means that if a field changes on the SAP SuccessFactors side, it then needs to be changed in the custom solution. All of the form processing happens outside of the SAP SuccessFactors solution, and the data is only sent to SAP SuccessFactors after the final form approval is complete.

Another approach is to dynamically read both the requisition template attributes and the user permissions to build the form. Fortunately, SAP provides APIs to retrieve both the fields of the requisition template as metadata and the permissions for a specific user. Figure 9-2 shows the relationship between the requisition template, the field controls (permissions), and the user (hiring manager or approver).

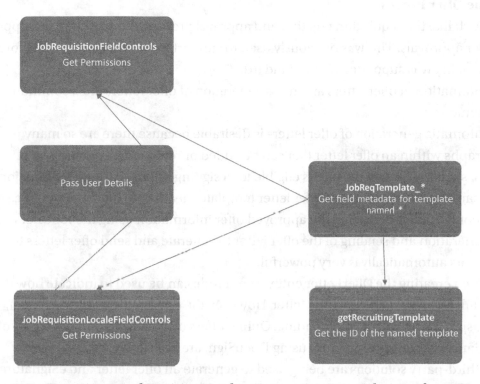

Figure 9-2. *Functions and Entities Used to Determine Req Behavior for a User*

Extracting Application Question Responses Using Integration Center

Reporting on requisition-specific application questions is a common request that is not supported by the standard ad hoc reporting. Where such reports are needed, it's common to have a daily Integration Center .CSV extract of Job Applications by navigating from the entity JobApplication to jobApplicationQuestionResponse.

As recruiters will not typically have access to the FTP folder that the question responses are stored in, it is also common to copy the resultant file to a Sharepoint folder that recruiters have access to.

Extracting Interview Assessments Using Integration Center

This is a similar reporting request to the application question response scenario earlier and employs the same solution, with the JobApplicationInterview entity (the interview) and from there to the entities for InterviewOverallAssessment and InterviewIndividualAssessment.

The Offer Process

Much like the requisition creation and approval process, the same can be supported for offer approvals. This was previously used to support mass offer approvals before that functionality was supported in the standard.

Automation of offer letters and mass generation of offer letters are a common requirement.

Automatic generation of offer letters is desirable because there are so many paragraphs within an offer letter that can be added or removed, depending on offer details, such as whether the hire is eligible for a signing bonus. Without automation, there can be a lot of variants of offer letter templates, leading to high maintenance effort.

A combination of reading the approved offer information using the JobOffer entity and generation and sending of the offer letter to generate and send offer letters to applicants automatically is very powerful.

When creating the OfferLetter entity, sendMode can be used to indicate how the applicant should receive the offer letter. However, the options are limited to email, emailaspdf, print, verbal, and pending. Online offers using SuccessFactors native offer acceptance, as well as eSignature using DocuSign, are not supported.

If third-party solutions are being used to generate an offer letter and eSignatures are to be leveraged, then the sendMode should be "pending." The recruiter can then see the generated offer and select the appropriate method, including eSignature.

An alternative approach is to define offer letter paragraphs as token fields within the offer approval that are populated using business rules.

For mass generation of offer letters, for large groups of similar hires, such as interns, a third-party document generator may be appropriate. SuccessFactors can be used to generate the offer letter using the OfferLetter entity and then the copy of the letter sent to the third-party solution.

Further Tips on Recruiting ODATA Entities

In addition to the aforementioned reference guides, a good place to look for up-to-date ODATA entity information is the SAP API Business Hub. This can be found at `https://api.sap.com`.

Choose *SuccessFactors* ➤ *APIs* ➤ *Packages* ➤ *SAP SuccessFactors Recruiting*.
As you can see in Figure 9-3, you can then navigate to the individual entities.

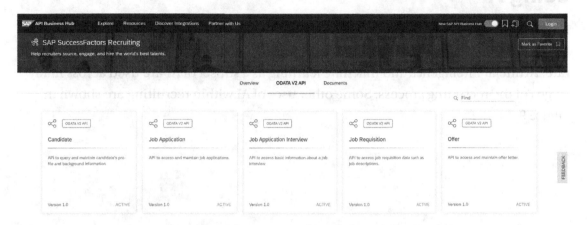

Figure 9-3. *Recruiting Entities Available in the SAP API Business Hub*

From there, you can look at the data model, documentation, or even try out the
API. As you can see in Figure 9-4, you can even see code snippets to access the entities.

Figure 9-4. *A Sandbox Environment Is Available to Try Out the ODATA
Entities*

With the resources that SAP have provided for integration, the sky is the limit.

213

Case Study: Matching Applications to Requisitions Using AI

In Chapter 5, we saw the use of AI chatbots to engage with site visitors, thus creating a better candidate experience and increasing the site visitor-to-application conversion rate. Artificial intelligence and machine learning as a subset of AI are used across many stages of the recruiting process. Some other uses of AI within recruiting are shown in Figure 9-5.

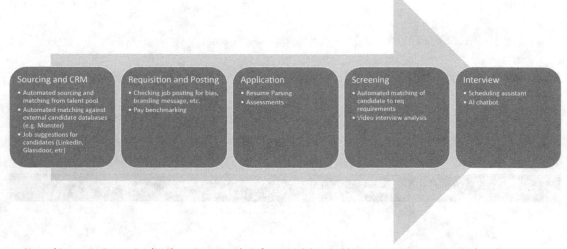

Natural Language Processing (NLP) sentiment analysis from candidate and hiring manager surveys, social media, etc.

Figure 9-5. *Use of Artificial Intelligence Across the Recruiting Process*

Of course, AI can do so much more, including offering competitiveness analysis, onboarding chatbot assistance, mentor matching, and so on. The sky is the limit!

Buyer Beware: Machine learning is only as good as the data that has been used to train it. Recruiting is ripe for bias. AI can both reduce unconscious bias and increase it if the learning data has embedded biases.

At the time of writing, SAP recently announced the acquisition of SwoopTalent, an AI-powered platform for the talent life cycle. This will have a huge impact on the capabilities around the employee experience, external candidate engagement, and organizational talent analytics. Example uses for employees include learning recommendations, career development, internal projects, and job matching. Talent intelligence is the new buzzword! The initial focus of the SwoopTalent platform seems to be on internal mobility and analytics, but it's inevitable that there will be an impact on the recruiting offering.

A common recruiting use case for AI is the matching of job requisitions to candidates. This may be for the following:

- Sourcing and matching of talent that have given their information to other recruiting sites such as Monster.com. This may be against active requisitions.

- Sourcing and matching of talent that are already in the talent database: employees, previous applicants, etc.

- Matching and ranking of candidates that have applied against a specific requisition.

For our case study, we will use the last example. The business case for automatically ranking applications against a job requisition is easy. Recruiters are often required to own many job requisitions, which in turn may have many applications. Sifting through the morass of applications, recruiters spend more of their time on repetitive resume review, leaving less time to effectively communicate with the right candidates, as well as hiring managers.

We will use a third-party solution to automatically rank applications against job requisitions. As with previous examples, we'll show example use of the APIs on the SAP SuccessFactors side without recommending a particular solution.

We have the following assumptions when performing automated matching:

- We want to use Job Profile Builder to ensure consistent requirements and job descriptions. Our job descriptions have already been scanned by a third-party tool to ensure they're on brand and don't include gender biases.

- Our organization hasn't validated the matching of applications against all our job roles. For this reason, we will only match if the requisition is tied to certain job roles. We're told that the matching solution will handle this.

- We want recruiters to give feedback on the matching results, so that the matching engine can learn.

As can be seen in Figure 9-6, the steps to be performed are

1) Read open job requisitions that have a relevant job role, including job profile with description, competencies, etc.

2) Read relevant applications for this job requisition and their resume, background information, etc. They may be filtered by their status (e.g., only new applications) and whether the third-party matching engine has already provided a matching score.

3) Generate a matching score for each application in the third-party matching engine.

4) Update the SAP SuccessFactors application with the matching score.

5) If applicable, the recruiter gives feedback by clicking a link to the matching solution.

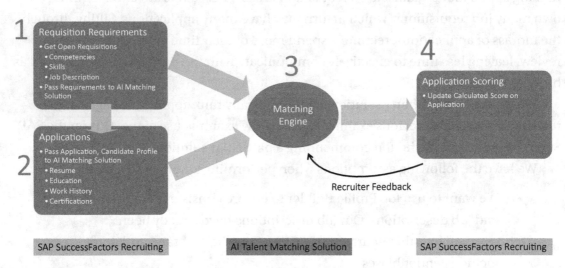

Figure 9-6. *Steps in Our Automatic Matching Example*

We will need a couple of additional application fields that will be populated by the matching engine and are defined as read-only for the recruiter. These are

- **Job Requirement Matching Score** – This will be a percentage.

- **Matching Info and Feedback** – This will be a hyperlink that takes the recruiter to the matching app using single sign-on (SSO). They will be able to then give feedback on the score.

Figure 9-7 shows an example matching field definition within the application template. This is can be found in *Manage Templates* ➤ *Recruiting Management* ➤ *Job Application*. Then choose the appropriate application template. It may be that automated matching is not used on all application templates. For example, it may not be suitable for manufacturing shop floor jobs where resumes are not common and may not be required on the application. Note that Forward Intact should be changed to False as the matching score and subsequent results link should be done separately for each requisition and application.

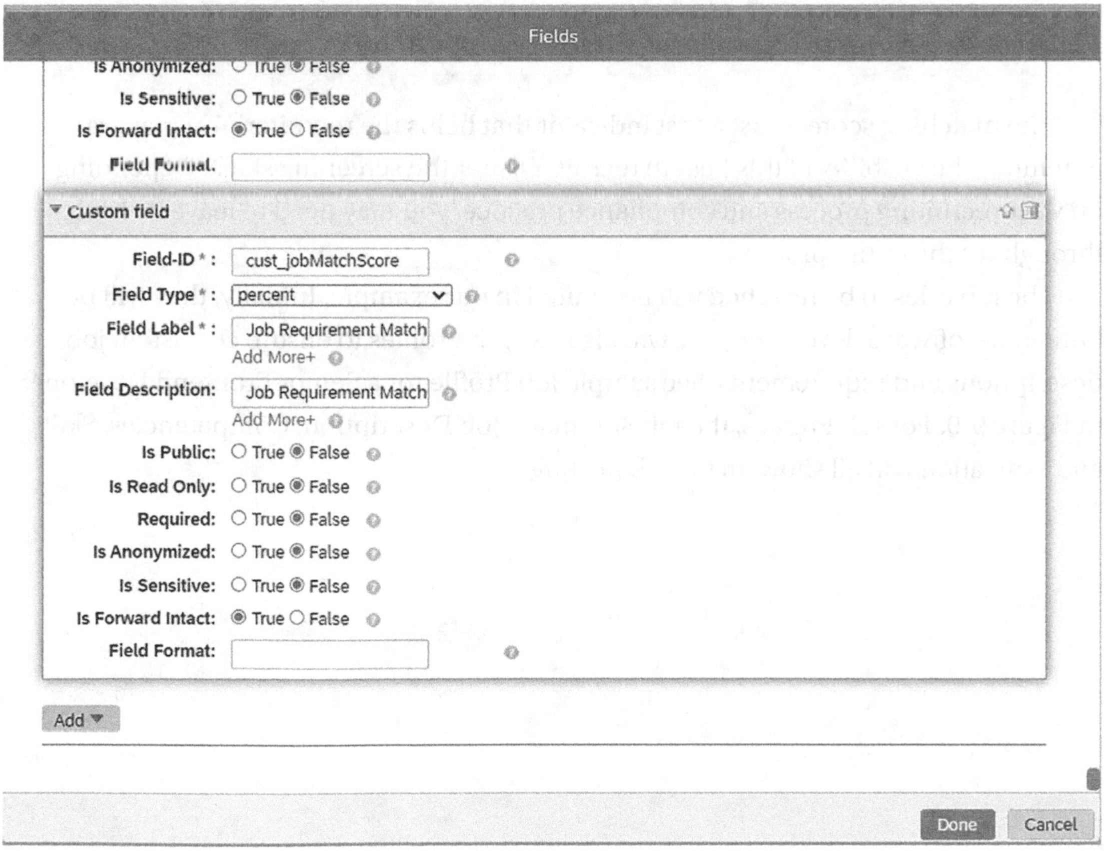

Figure 9-7. *Example Matching Field Definition*

We will also add an application field of type hyperlink. It will be also populated by the matching solution and will be a deep link to the matching results within the matching solution. See the recruiters view in Figure 9-8.

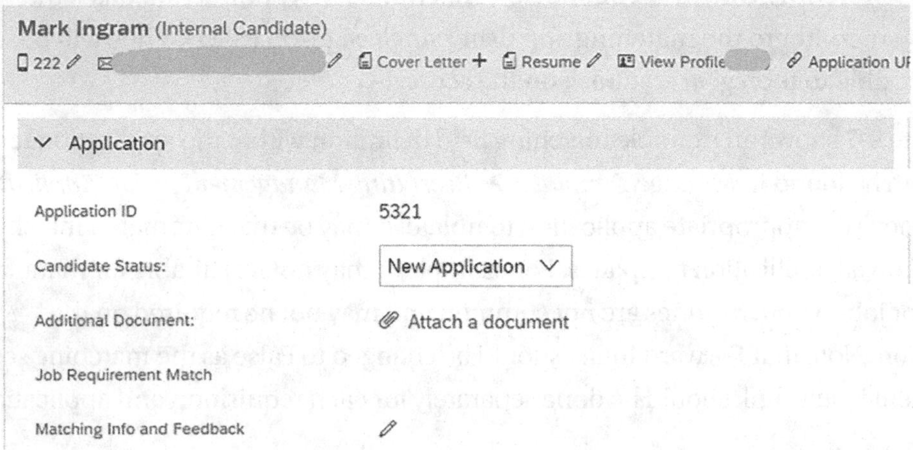

Mark Ingram (Internal Candidate)

222 ✏️ ✉️ ▇▇▇▇▇▇▇ ✏️   📄 Cover Letter ➕ 📄 Resume ✏️ 📖 View Profile ▇▇ 🔗 Application UF

∨ Application

Application ID	5321
Candidate Status:	New Application ∨
Additional Document:	📎 Attach a document
Job Requirement Match	
Matching Info and Feedback	✏️

Figure 9-8. *Recruiter View of Matching Score on an Application.*

The matching score is just a first indicator that helps the recruiter. You may want to remove the visibility of this field to recruiters after the screening stage. Depending on your recruiting process and compliance practice, you may need to leave it visible throughout the entire process.

The job roles to be matched will be limited in our example. Initially, they will be limited to software developers. We will also use Job Profiles to ensure consistent job descriptions and requirements. See sample Job Profile for a Senior Frontend Developer in Figure 9-9. For Job Profiles, the Job Summary, Job Description, Competencies, Skills, and Education will all show on the job posting.

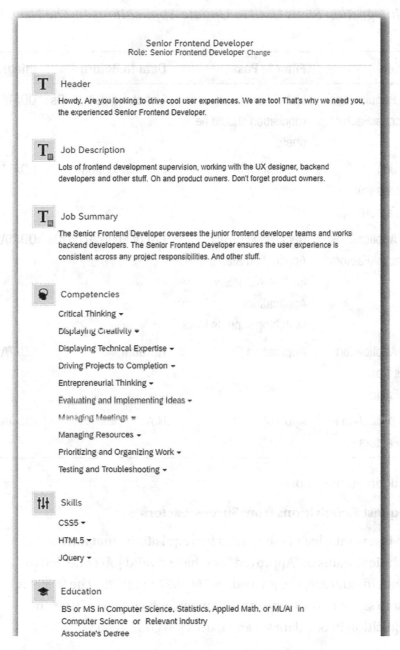

Figure 9-9. *Example Job Profile for Senior Frontend Developer*

Let's begin by breaking down the integration points between SAP SuccessFactors Recruiting and the matching solution. Table 9-1 breaks the integration down into five steps. Note that some steps can be combined. For example, job descriptions can be returned with requisition IDs. It depends on the needs of the matching solution.

Table 9-1. *Integration Steps Used to Update Applications In The New Application Status*

Step	Integration	Filter to Pass	Data to Return	Integration Type
1	Request Requisitions from SuccessFactors	List of job codes, requisition should be open.	List of requisition IDs	ODATA Get Query
2	Request Job Descriptions from SuccessFactors	Requisition ID	Job description	ODATA Get Query
3	Request Applications from SuccessFactors	Requisition ID, Job Application Status is "Default" (New Application). Matching score is blank.	List of application IDs	ODATA Get Query
4	Request Application Resumes from SuccessFactors	Application ID	Resume	ODATA Get Query
5	Update Application in SuccessFactors	Application ID	N/A	ODATA Upsert

Let's break down the steps:

1) **Request Requisitions from SuccessFactors**

   We just want a list of job req IDs for requisitions that have a technical status of "Approved" and have a valid job code tied to them. In our case, the job code is 7000027 or 99999. The 99999 is just to show how to pass more than one job code. We have one requisition in our data set and it uses job code 700027.

   Using the neat trick from Chapter 8, we can create an outbound integration, export the integration specification, and get the last line from the downloaded file. This contains the URL to make the call:

   ```
   /odata/v2/JobRequisition?$select=jobReqId&$filter=(
   internalStatus eq 'Approved' and jobCode in '7000027',
   '99999')
   ```

With the experience that we've gained in previous chapters, we can also put the URL directly into Postman using a GET request. We would see the result in Figure 9-10.

Figure 9-10. *A GET Operation to Retrieve Open Requisitions for Certain Job Codes, Shown in Postman*

Yay! We have returned our one job requisition for a senior front-end developer.

2) **Request Job Descriptions from SuccessFactors**

Let's get the job descriptions (or single job description in this example). We are using US English as the job description locale. Note that we could have returned the job description with the job req IDs. Either way is possible, and it depends on how the third-party matching solution needs to integrate.

If we requested all of the relevant job descriptions and job req IDs at the same time, we would formulate the GET request something like this:

```
/JobRequisition?$select=jobReqId,jobReqLocale/externalJob
Description&$expand=jobReqLocale&$filter=(internalStatus eq
'Approved' and jobCode in '7000027','99999')&$format=json
```

We have built on the original request by changing the $select parameter
to include jobReqLocale/externalJobDescription. We have also added an
$expand parameter of $jobReqLocale. The $expand parameter instructs the
query to read jobReqLocale by navigating from the jobRequisition entity.

Figure 9-11 shows the GET request as well as the results in Postman.

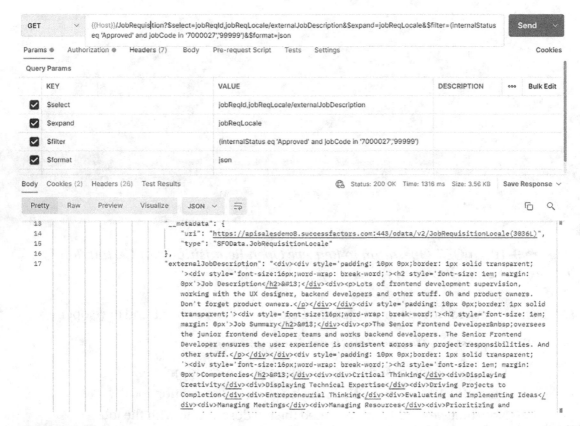

Figure 9-11. *The Results of a GET Operation to Retrieve a Job Description, Shown
in Postman*

3) **Applications from SuccessFactors**

Let's grab some applications before the recruiter gets impatient
and starts screening them.

The third-party matching solution will pass one or more job requisition IDs. A list of application IDs and resumes will then be returned. Note the following:

- The full structured extract of the application and candidate profile may be also passed to the matching solution. We are providing just the resume as an example.

- The resume will need to be extracted as a second call to SAP SuccessFactors. The reason is that documents aren't directly stored on the job application or candidate profile. Only a document reference is stored.

In Figure 9-12, we can see a sample of the data that we are pulling using Integration Center. We don't need to build an Integration Center integration. This is shown for illustration purposes.

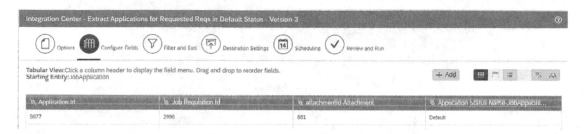

Figure 9-12. *Extracting Attachment IDs from Applications in Integration Center*

Figure 9-13 shows the filters that we will be using to pull job applications. We are simply passing a list of job requisition IDs and specifying that the job application status should be "Default," meaning it's a new application. Listing the Application Status Name as a column isn't necessary. We wanted to just check it for correctness.

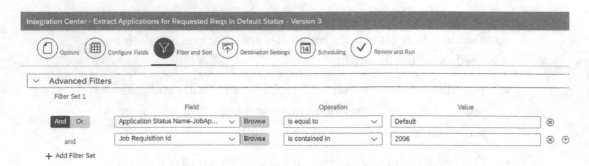

Figure 9-13. *Selecting Applications in Default Status in Integration Center*

If we extracted the ODATA request and removed the unnecessary application status, it would be

```
/odata/v2/JobApplication?$select=applicationId,jobReqId,resume/attachme
ntId&$expand=resume,jobAppStatus&$filter=(jobAppStatus/appStatusName eq
'Default' and jobReqId in '2996')&$format=JSON
```

The GET call and response can be seen in Postman in Figure 9-14.

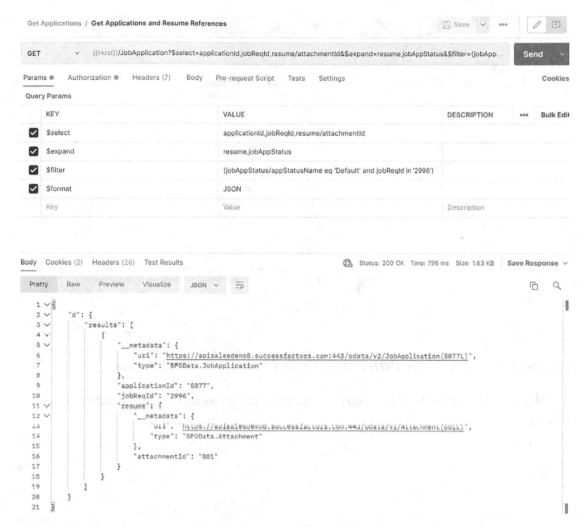

Figure 9-14. *The Results of Requesting Attachment IDs of Applications Using Postman*

The next step is to request the resumes. For each application ID and resume attachment provided, we need to request the contents of the attachment. The beauty is that the request for the attachment is already shown in line 13. Any resources that are requested via JSON will also show the URI to get that resource.

If you haven't tried to get attachments before using an integration user, you may be missing the permission "ODATA API Attachment Export." You can see the permission that you need in Figure 9-15.

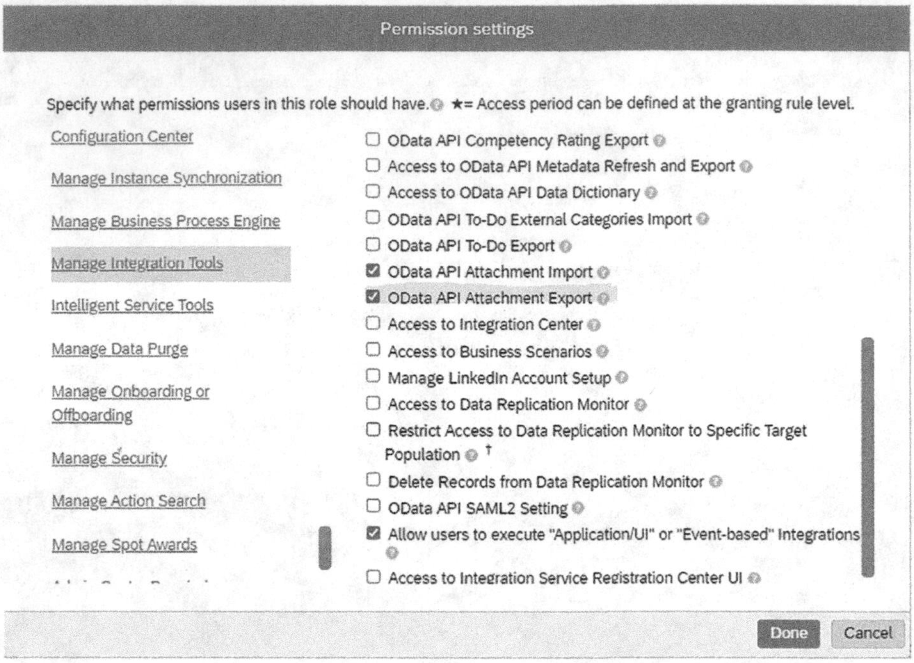

Figure 9-15. *Role-Based Permission Needed to Extract Attachments*

Figure 9-16 shows the call to retrieve the attachment using Postman, including the results.

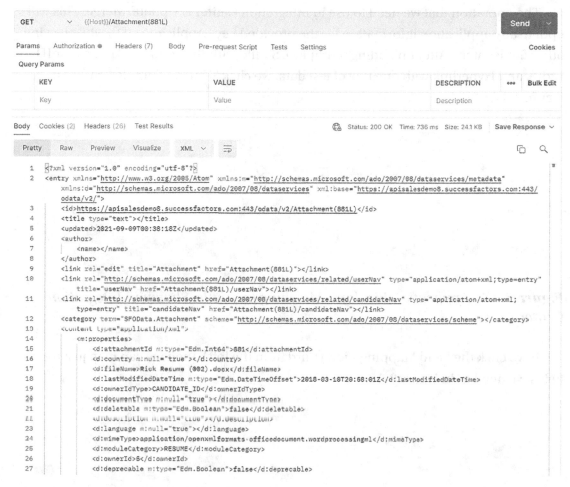

Figure 9-16. *Results of Requesting Attachment Using Postman*

The file content is not shown in the figure. It is binary form and doesn't make for interesting reading, unless you are an AI-based resume-to-requisition ranking system.

4)   Update Application in SuccessFactors

The work of calculating a score from a job description and resume is then done by the matching system. This will take many steps, the first of which may be removing any resume information that could cause bias in the ranking engine, such as name, address, age, and photo (in some countries).

Once we have that magic matching score, we will ask SuccessFactors to update it, along with a hyperlink field which will take the recruiter to the details of that matching in the matching engine. We will illustrate the integration in Integration Center as if we were using a file-based import. After that, we will show how to do it in Postman.

The operation that we need to use in Integration Center is Update/Merge. This will update any application information for the provided key (ApplicationID), while leaving other fields intact. After uploading a simple .CSV file with the columns ApplicationID, Score, and Hyperlink with one row of test data, we click "Upload Sample CSV" as in Figure 9-17.

Figure 9-17. *Application Fields for Matching Example, Shown in Integration Center*

If we click the Field Mapping View (third from the left), then we can map just the fields we need as in Figure 9-18.

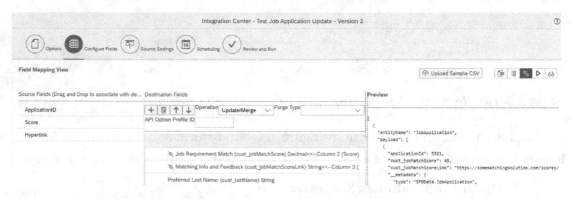

Figure 9-18. *Mapping of Application Fields and Resultant Preview of the Necessary Payload*

If we look at the Preview pane to the right, the payload section is a great example that we can cut and paste into Postman to test. You may want to keep it handy in a Notepad file.

Open the Postman desktop client application and create a new HTTP Request using the steps.

Start by creating a new HTTP Request as shown in Figure 9-19.

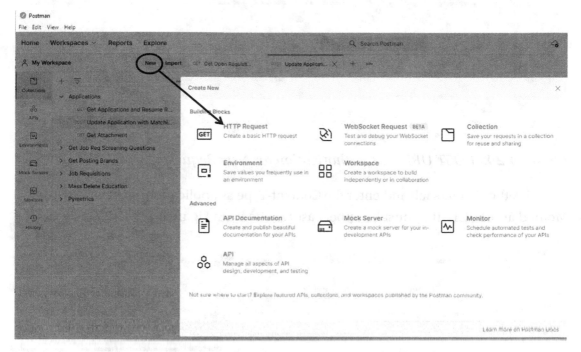

Figure 9-19. *Creating a New HTTP Request*

Change the request type from GET to Post and change the URL to *{{$host}}/ Application(5321)* before selecting the Authorization tab and selecting Basic, using {{username}} as the Username and {{password}} as the Password, as shown in Figure 9-20. This will then reuse the environment variables for your host, username, and password.

Note   Basic Authentication is just for simplification of examples. Do not use it in client environments.

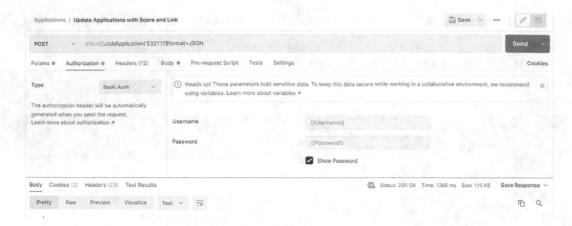

Figure 9-20. *POST URL and Authorization for New Request*

Click the Headers tab and enter the Content-Type as application/json and X-Http-Method as MERGE (this must be upper case) as shown in Figure 9-21.

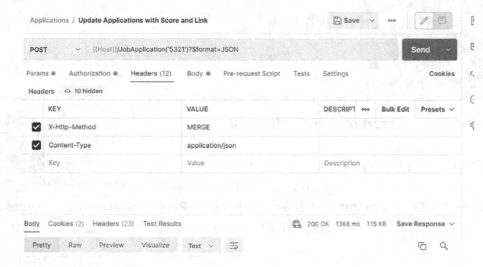

Figure 9-21. *Header Information for New Request*

We can now add the body of the request. We haven't done this before in Postman as we have just been performing GET requests and not POSTs. See the body in Figure 9-22.

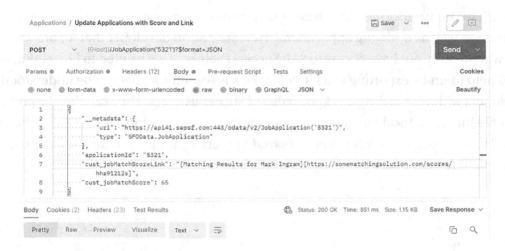

Figure 9-22. *Body of New Request*

Note the square brackets around the URL field for the matching results. The first pair of square brackets contains the hyperlink as it should be shown. The second contains the actual hyperlink. If the two are the same, then put the URL in both pairs of square brackets. The Knowledge Base Article (KBA) 3047601 from SAP explains it here https://launchpad.support.sap.com/#/notes/3047601.

You can see the updated application in Figure 9-23. They probably won't be getting the job.

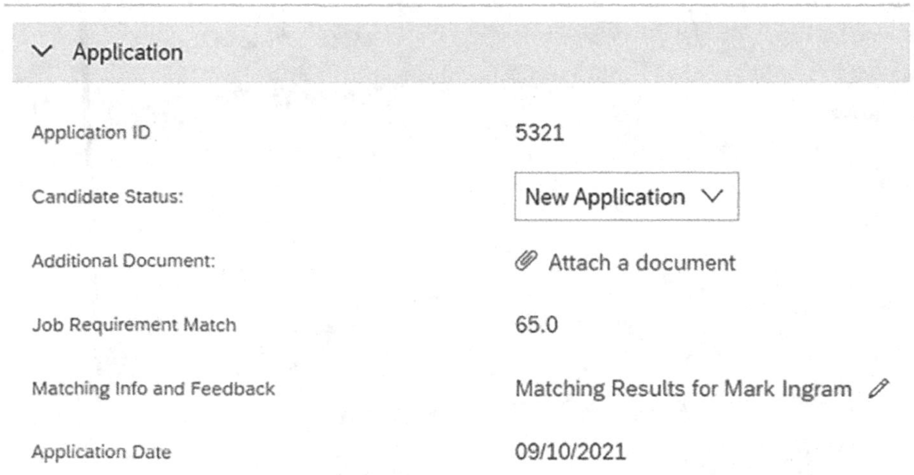

∨ Application	
Application ID	5321
Candidate Status:	New Application ∨
Additional Document:	📎 Attach a document
Job Requirement Match	65.0
Matching Info and Feedback	Matching Results for Mark Ingram ✎
Application Date	09/10/2021

Figure 9-23. *The Updated Application in SuccessFactors*

We have now performed all the necessary steps for reading job descriptions and resumes, as well as updating applications. The integration developer that we are working with will possibly not have access to SAP SuccessFactors. If we wanted to be helpful, we could help them by exporting to a JSON file. This is something that can be understood by developers without needing any knowledge of SuccessFactors. They can then import the JSON file into their local Postman to try out the API. You can export the JSON by clicking the "..." to the right of the collection you saved the service under and then select Export, as shown in Figure 9-24.

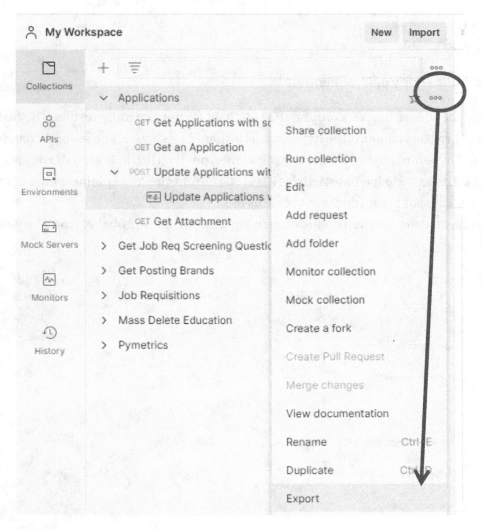

Figure 9-24. *Exporting the Collection of Requests as a JSON File*

Way to go! The exercises you have done here are applicable to other scenarios and ODATA entities.

Conclusions

You have now learned the following:

- How the ODATA APIs delivered by SAP SuccessFactors can support many different use cases, from reading requisition template configuration for custom UIs to supporting new candidate experiences.

- How artificial intelligence and machine learning can be used throughout the recruiting process.

- How to perform updates to job applications using MERGE using Integration Center in SAP SuccessFactors or Postman. This same information can be applied to other ODATA entities.

CHAPTER 10

Background Check Integration

Background checks are an integral part of any company's recruiting system. These checks are incorporated into the overall recruiting workflow and typically initiated just before the candidate is given a firm job offer, although some customers might prefer to conduct these checks just before the actual hire. Automating the background check integration greatly simplifies the overall recruiting process and frees up a recruiter's resource to focus on the hiring process.

SuccessFactors Recruiting system offers a framework for integrating third-party background check vendors using existing tools and technologies. The very first background check vendor integration with SuccessFactors integrated was First Advantage, and it still exists as part of the standard documentation. The industry has several big players in the United States, and their popularity is increasing globally especially in the Asia Pacific region. Imagine that a customer asks you to build an integration to a vendor. How would you approach it? In this chapter, we will use a background check vendor called MYTHICAL as we go through several steps.

General Notes

- Technically, multiple background check vendors can coexist in one customer instance; however, one of them, typically the incumbent, is designated as the default and the results will be visible in the background check portlet. All other vendors must be configured via custom fields on the job application.

- Background check packages must be manually defined on the job requisition at the time of creation.

235

© Anand 'Andy' Athanur, Mark Ingram and Michael A. Wellens 2022
A. A. Athanur et al., *Innovative SAP SuccessFactors Recruiting*, https://doi.org/10.1007/978-1-4842-7425-5_10

- The system can return results for only one package at a time. For example, if you have a package, say, for Employment Check as in the example, you can only return results for Employment Check. You cannot break this down to sub-requests such as a) current employment verification, b) previous employment verification, and c) address verification. It's up to you to configure the packages as you see fit for the customer.

SuccessFactors Recruiting offers you options for implementing the background check integration:

- **Option 1** – Via recruiter action in the UI and is referred to as APPUI going forward

- **Option 2** – Using Intelligent Services and is referred to as IS going forward.

- Table 10-1 lists the two options and their setup.

Differences Between Option 1 (APPUI) and Option 2 (IS)

Table 10-1. *Comparison of APPUI and Intelligent Service Triggers for Background Checks*

Item	Option 1	Option 2	Notes (if applicable)
How to enable integration	Admin Center ➤ Background Check Central	Intelligent Services	
How to initiate	UI via Take Action ➤ Initiate Background Check	Changing Job Application Status to "Background Check"	The actual name of the status is configured in the job requisition template
Intelligent Services Event	N/A	Update of Job Application Status	
Integration Center template (ICD)	A starter ICD can be downloaded from the SAP API Business Hub		A sample template can be downloaded from https://api.sap.com/search?filter=(Type%20eq%20(ICD))
Starting Entity in ICD	JobApplicationBackgroundCheckRequest	JobApplication	
Association of ICD template	Via Admin Center ➤ Manage Data ➤ Recruiting Vendor Integration Mapping	Via Admin Center ➤ Intelligent Services Center ➤ Update of Job Application Status (Event) ➤ Integration	

(continued)

237

Table 10-1. (*continued*)

Item	Option 1	Option 2	Notes (if applicable)
Number of simultaneous background check requests	1	Multiple	Multiple events raised and viewable within Intelligent Services Center
Filtering possible in ICD? Yes/No	No	Yes	Filter is based on the name of application status configured for background check (varies by configuration)
JSON Response field mapping required? Yes/ No	Yes. Expected for entity JobApplicationBackgroundCheckRequest for filling in vendor order number, order status, and other fields	No. Vendor is expected to create a record for JobApplicationBackgroundCheckRequest with vendor order number, order status, etc.	
Integration Trigger	triggerType="APPUI"	triggerType="IS"	

Step 1: Permissions in Admin Tools – Enable Background Check Integration

Log into the customer's instance as an administrator and navigate to Admin Tools ➤ Manage Permission Roles. Choose System Admin as the Permission Role and ensure the following are selected. The SuccessFactors Recruiting system requires several permissions to be set up for different roles. Table 10-2 shows the applicable list of permissions.

Table 10-2. *Role Based Permissions Needed for Background Checks*

System Role	Permission Category	List of Permissions
System Administrator	Manage Recruiting	– Background Check Central
System Administrator	Metadata Framework	– Admin Access to MDF OData API – Manage Data
Recruiter	Recruiting Permissions	– Background Check Initiate Permission – Background Check Update Permission
API User	General User Permissions	– User Login – SFAPI User Login
API User	Recruiting Permissions	– Background Check Initiate Permission – Background Check Update Permission – ODATA API Application Create – ODATA API Application Export – ODATA API Application Audit Export – ODATA API Application Update

(continued)

Table 10-2. (*continued*)

System Role	Permission Category	List of Permissions
API User	Manage Integration Tools	– Access to Event Notification Subscription – Access to Event Notification Audit Log – Manage OAuth2 Client Applications – Access to ODATA API Audit Log – Manage ODATA API Basic Authentication – Access to API Center – Access to ODATA API Metadata Refresh and Export – Access to ODATA API Data Dictionary – Access to Integration Center – Allow users to execute "Application/UI" or "Event-based" Integrations
API User	Intelligent Service Tools	– Intelligent Services Center (ISC)

Step 2: Create Picklists for Background Check

The background check integration is predicated on the vendor defining packages for each type of check based on the job requisition template. See Figure 10-1 for such a picklist definition.

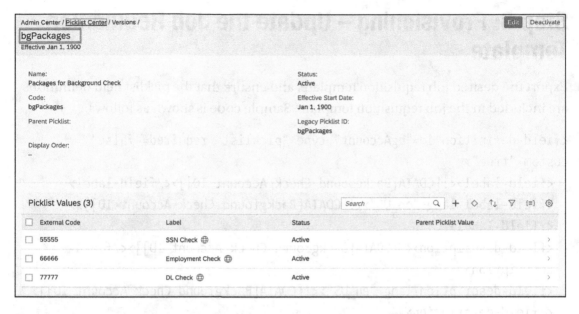

Figure 10-1. *Package Definition for a Background Package*

Some background check vendors may require an optional configuration for associating accounts depending on company location, department, or division. Shown in Figure 10-2 is a sample picklist definition for such an account.

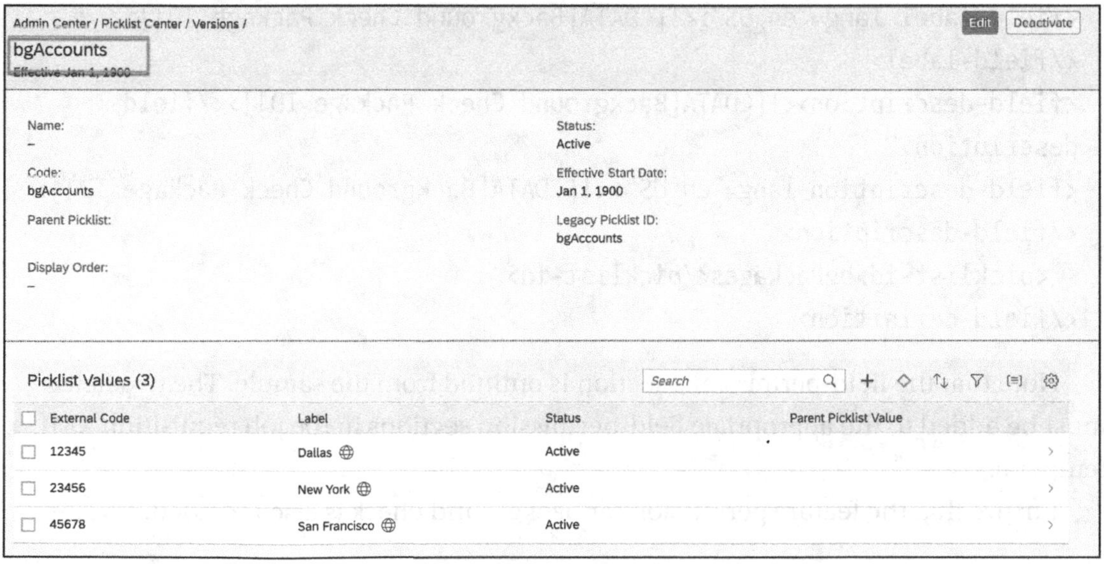

Figure 10-2. *Sample Picklist for Defining an Account Code*

Step 3: Provisioning – Update the Job Requisition Template

Export the desired job requisition template, and ensure that the picklist field definitions are included in the job requisition template. Sample code is shown as follows:

```
<field-definition id="bgAccount" type="picklist" required="false"
custom="true">
  <field-label><![CDATA[Background Check Account ID]]></field-label>
  <field-label lang="en_US"><![CDATA[Background Check Account ID]]>
  </field-label>
  <field-description><![CDATA[Background Check Account ID]]></field-
  description>
  <field-description lang="en_US"><![CDATA[Background Check Account ID]]>
  </field-description>
    <picklist-id>bgAccounts</picklist-id>
</field-definition>
<field-definition id="bgPackage" type="picklist" required="false"
custom="true">
  <field-label><![CDATA[Background Check Package ID]]></field-label>
  <field-label lang="en_US"><![CDATA[Background Check Package ID]]>
  </field-label>
  <field-description><![CDATA[Background Check Package ID]]></field-
  description>
  <field-description lang="en_US"><![CDATA[Background Check Package ID]]>
  </field-description>
    <picklist-id>bgPackages</picklist-id>
</field-definition>
```

Note that the field-permissions section is omitted from the sample. The two fields must be added to the appropriate field-permission sections in the job requisition XML template.

Ensure that the feature permission for background check is also included.

Sample code is as follows:

```
<feature-permission type="backgroundCheck">
    <description><![CDATA[The following roles can launch background
    check]]></description>
    <role-name><![CDATA[S]]></role-name>
    <role-name><![CDATA[O]]></role-name>
    <role-name><![CDATA[R]]></role-name>
    <role-name><![CDATA[G]]></role-name>
    <role-name><![CDATA[V]]></role-name>
    <role-name><![CDATA[Q]]></role-name>
    <status><![CDATA[Background]]></status>
</feature-permission>
```

Note The status name defined here "Background" must match the status name on the Applicant Status configuration. See Figure 10-3.

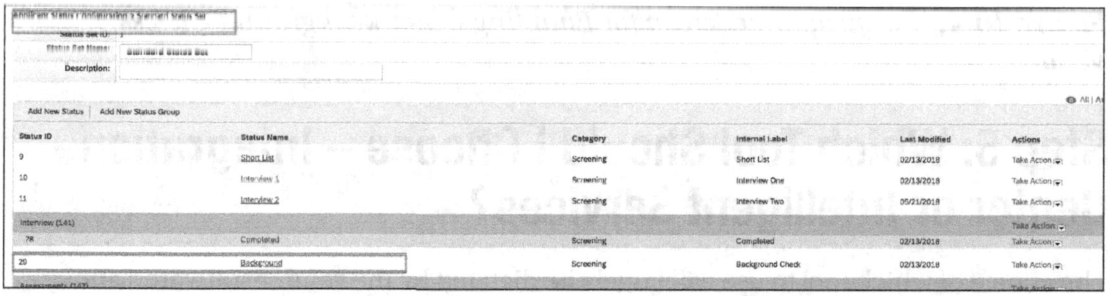

Figure 10-3. Applicant Status Configuration Showing the Background Check Details

Once you make the desired changes in your job requisition template, import them back into Provisioning and ensure that there are no errors in your template. Warnings in green are informational. You may choose to correct them.

Step 4: Admin Tools – Enable Other Background Check Vendor

Navigate to Admin Center ➤ Background Check Central, and ensure that your screen resembles Figure 10-4.

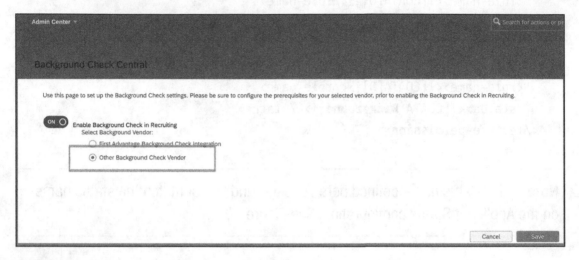

Figure 10-4. *Confirmation Screen for Enabling Other Background Check Vendor*

Step 5: Which Tool Should I Choose – Integration Center or Intelligent Services?

The choice of which tool to use will largely be dictated by the vendor software's ability to support JSON/XML or whether it's SOAP based. If the vendor can support JSON/XML, Integration Center will be an excellent choice. However, if the external system can only support SOAP, then the option will be Intelligent Services. The rest of the chapter goes into considerable detail on using Integration Center to build your integration. For Intelligent Services, you have to rely on the standard available documentation on help.sap.com.

Step 6: Which ICD to Use – Option 1 (APPUI) or Option 2 (IS)?

In the beginning of this chapter, we described what are the similarities and differences between the two options. You should consult with your customer which option they would like to use. For a majority of the cases, customers will choose Option 1 (APPUI), which means that a recruiter will manually initiate a background check in the job application UI. In some cases, a customer may already have an incumbent background check vendor in place. In such situations, Option 2 (IS) will be a good option. What this means is that a background check will be initiated by the system when the job application reaches a designated status in the applicant workflow.

Regardless of which option you choose, the end result is that you can integrate a background check vendor in the recruiting process.

Step 7: Setting Up the Integration Center Template (ICD) for Option 1

Access Integration Center and navigate to My Integrations. Your screenshot should resemble Figure 10-5.

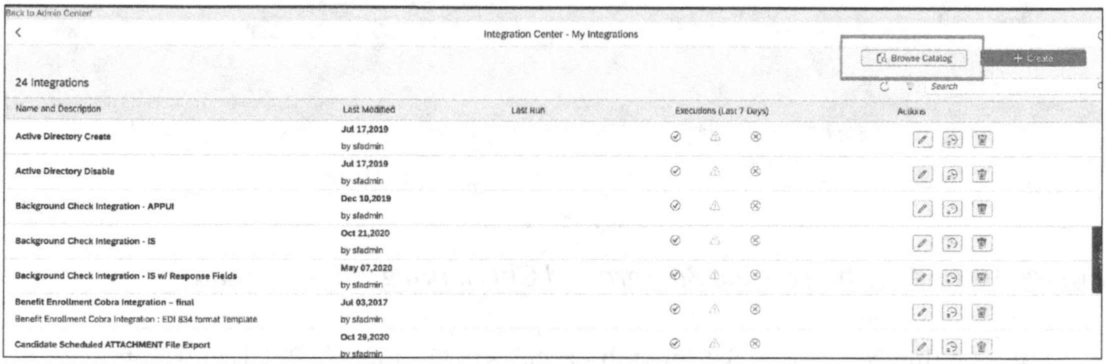

Figure 10-5. *Integration Center ➤ My Integrations*

Now click "Browse Catalog" as shown in Figure 10-5 and locate the template called "SuccessFactors Integration Center – Integration Template for the Recruiting Background Check Functionality (RCM)." The direct link is `https://api.sap.com/ package/SuccessFactorsIntegrationCenterTemplatefortheRecruitingBackground CheckFunctionalityRCM?section=Overview`. Note that you will require either an S-User ID established as a valid partner user or via self-registration at `https:// community.sap.com`. You will download the Integration Center Definition (ICD) file to your SuccessFactors tenant. Alternatively, you may go through Chapter 7 and create your own ICD. Ensure that your starting entity in Integration Center is JobApplicationBackgroundCheckRequest. A sample of the starting entity is shown in Figure 10-6.

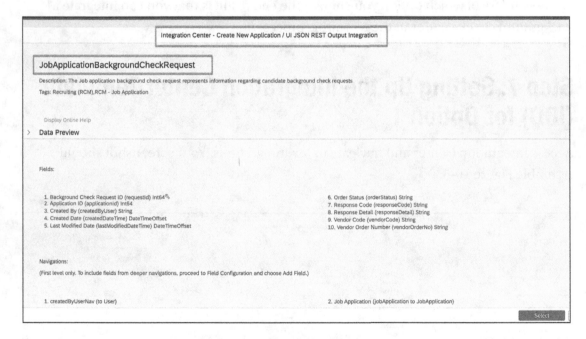

Figure 10-6.   *Creating a New Background Check Integration Manually*

Notes on the Destination Settings tab – applies to Option 1 (APPUI) and Option 2 (IS):

- The REST server settings can be individually set for background check integration or use common settings from Security Center.

- The authentication options are none, Basic or OAuth (with SAML Bearer Assertion).

- If the authentication is none, then the vendor's software should inspect the payload to determine how it is handled.

- Basic Authentication is not recommended and will be disabled completely by the end of 2022.

Figure 10-7 shows the Destination Settings tab.

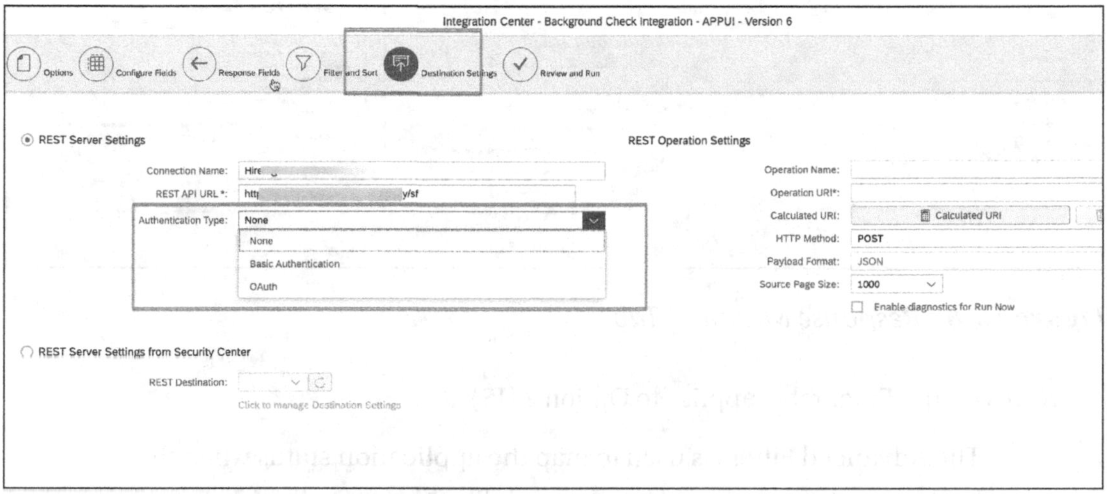

Figure 10-7. Destination Settings in Integration Center

Notes on the Response Fields mapping – applies to Option 1 (APPUI):

- backgroundCheckRequestOrderStatus, backgroundCheckRequest-VendorOrderNumber, backgroundCheckRequestResponseCode, and backgroundCheckRequestResponseDetail should be mapped within the Response Fields tab.

- The Operation should be set as Upsert Single.

- If the vendor's software is unable to produce a synchronous response, then the option available is to upsert a record manually into the JobApplciationBackgroundCheckRequest.

- If the response is delayed more than 60 seconds, the system will record a timeout failure and the background check must be reinitiated.

Figure 10-8 shows the completed sample response fields.

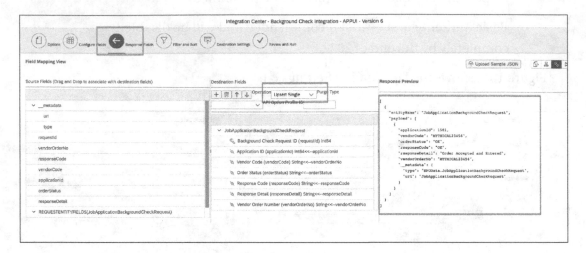

Figure 10-8. *Response Mapping Tab*

Notes on the Filter tab – applies to Option 2 (IS):

- The Advanced Filters is used to map the application status when the background check should be triggered. Please refer to Step 3 on how it's defined.

- You could use the status ID instead of the status name as long as you're sure the values are consistent.

- If your recruiting configuration has multiple templates, the filter status should account for all templates.

Figure 10-9 shows the completed Filter tab.

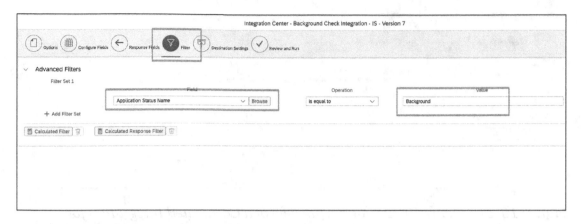

Figure 10-9. *Filter Tab Sample for Option 2 (IS)*

We will not go through the steps involved in creating and completing the integration template, but the final result should resemble Figure 10-10 for Option 1 (APPUI).

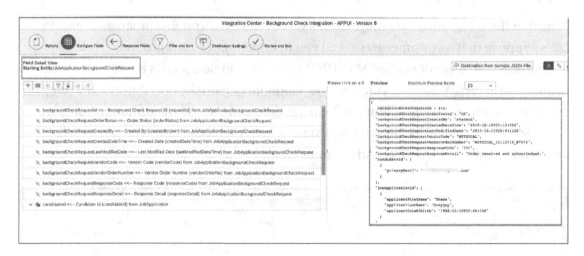

Figure 10-10. *Preview of a Completed Integration for Option 1 (APPUI).*

Shown in Figure 10-11 is a sample completed ICD for Option 2 (IS).

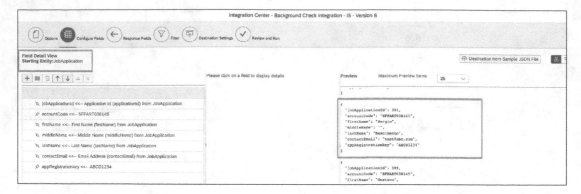

Figure 10-11. *Figure 10-12 shows a preview of a Completed Integration for Option 2 (IS)*

Step 8: Admin Tools – Define Recruiting Vendor, Applies to Option 1 (APPUI) and Option 2 (IS)

Now that the integrations are more or less set up, let's go through a practical scenario. Let's suppose that a customer has asked you to implement a background check integration with a vendor called MYTHICAL. The first logical step is to define the vendor in SuccessFactors. Assuming you have the adequate permissions as described in Step 1. You will navigate to Admin Center ➤ Manage Data and choose the option to "Create New." Scroll down or select Recruiting External Vendor. Please note that this step is common to Option 1 (APPUI) and Option 2 (IS).

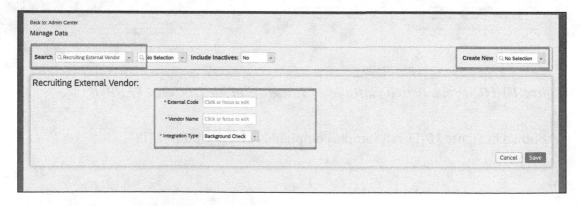

Figure 10-12. *Defining a New Background Check Vendor*

The completed Recruiting External Vendor will resemble Figure 10-13. If you have made an error, you can always click the Take Action ➤ Make Correction.

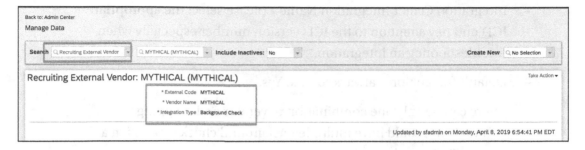

Figure 10-13. *Completed Background Check Vendor Definition.*

Step 9: Admin Tools – Associate Integration Center Template with Vendor, Applies to Option 1 (APPUI) and Option 2 (IS)

Similar to Step 8, you will navigate to Admin Center ➤ Manage Data, and click Create New and select Recruiting Vendor Integration Mapping. Your page will resemble Figure 10-14.

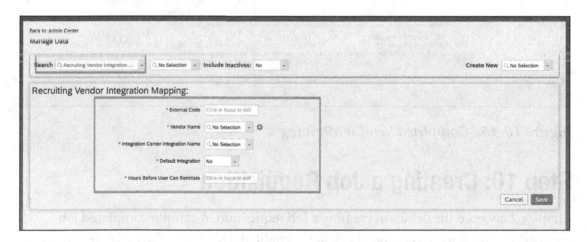

Figure 10-14. *Associating an Integration Center Template*

Notes:

- External code can be anything, but we suggest that you keep it the same as the vendor name.

- Vendor name will be shown in the drop-down.

- Integration Center Integration Name – please select the appropriate ICD and pay attention to the ICD version number especially when troubleshooting an integration.

- Default Integration – always set it to Yes.

- There can be only one combination of vendor and mapping, although you could have multiple background check vendors in a SuccessFactors tenant.

- Hours Before User Can Reinitiate has no real function at the time of this publication.

A completed mapping is shown in Figure 10-15.

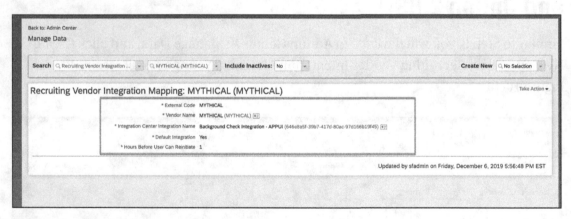

Figure 10-15.   *Completed Vendor Mapping*

Step 10: Creating a Job Requisition

Chapter 2 covered the details of creating a job requisition. A sample completed job requisition with the background check fields enabled is shown in Figure 10-16. Please note that the field "Background Check Account ID" is an optional field and has no effect on the actual background check.

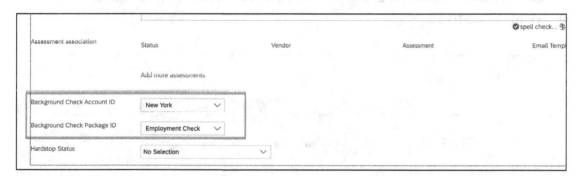

Figure 10-16. *Sample Completed Job Requisition with Background Check Fields Chosen*

Step 11: Recruiter Actions – Initiate Background Check

Login as the recruiter or proxy and navigate to Recruiting ➤ Job Requisitions. Select the job requisition you want to work on and move (or advance) the candidate to the Background Check status. The option to move or advance is dependent on the customer configuration. A note on the screenshots is shown – they are taken from different job requisitions and candidate, but the concept holds regardless of which candidate and job requisition combination you work with.

Figure 10-17 shows a sample job requisition with two candidates who are in the Background Check status.

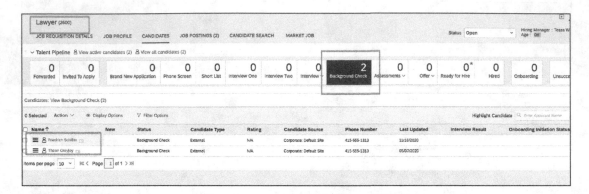

Figure 10-17. *Job Requisition with Two Candidates in the Background Check Status*

Click a desired candidate and go to Take Action ➤ Initiate Background Check as shown in Figure 10-18.

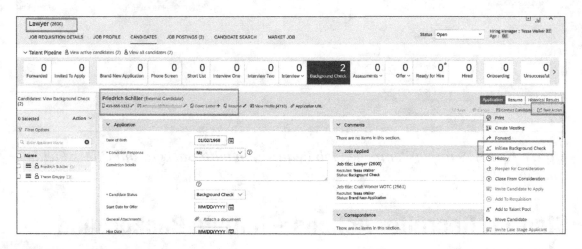

Figure 10-18. *Screenshot Showing How to Initiate a Background Check*

If the background check submission is successful, you should see a screenshot as in Figure 10-19.

Figure 10-19. *Sample Screenshot Showing Successful Initiation of a Background Check*

If you have made it this far, congratulations. You have successfully defined MYTHICAL as a background check vendor and have been able to initiate a background check from the Recruiting Job Application UI, namely, Option 1. We will examine Option 2 (IS) shortly.

But what happens if there is an error showing that your background check failed?

Step 12: Integration Center – Checking Results of Outbound Execution

Navigate to Admin Center ➤ Integration Center and click Monitor Integrations as shown in Figure 10-20.

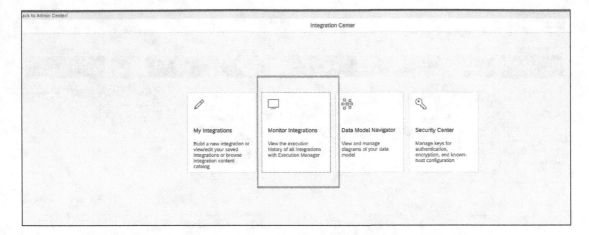

Figure 10-20. *Integration Center Highlighting the Integration Monitor*

The Execution Manager screen will open in a new pop-up where you can see the results of current and past executions. By default, the Execution Manager will show the executions in the last seven days. The filters can be adjusted depending on the time filter and optional the Process Status. An example is shown in Figure 10-21.

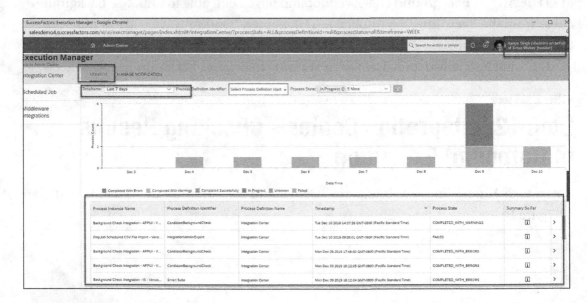

Figure 10-21. *Screenshot of the Execution Manager Showing Summary of Executions in the Last Seven Days*

Figure 10-22 shows the drill-down details of the first execution shown in Figure 10-21. To get the details, click the arrow and view the screen.

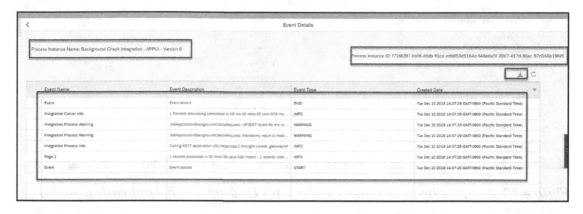

Figure 10-22. *Illustration of a Successful Background Check Request*

Detailed documentation on the Execution Manager is available on the standard Help Portal.

Notes:

1. This should be the starting point for troubleshooting any integrations.

2. Customers will often use the download link to export a CSV file for support issues.

3. You (or the customer) can optionally choose to notify an email address or an inbox by clicking Manage Notifications.

4. If the customer is initiating background checks in the user interface, you should see only one record per background check.

Step 13: Setting Up the Integration Using Intelligent Services Center (Option 2)

In the beginning of this chapter, we described two ways to integrate background checks, namely, via the UI (Option 1) and via Intelligent Services (Option 2). Chapter 12 of this book goes into some detail on how to use Intelligent Services. In this step, we will highlight how to set up the integration. Figure 10-23 shows a screenshot with the Update of Job Application Status highlighted.

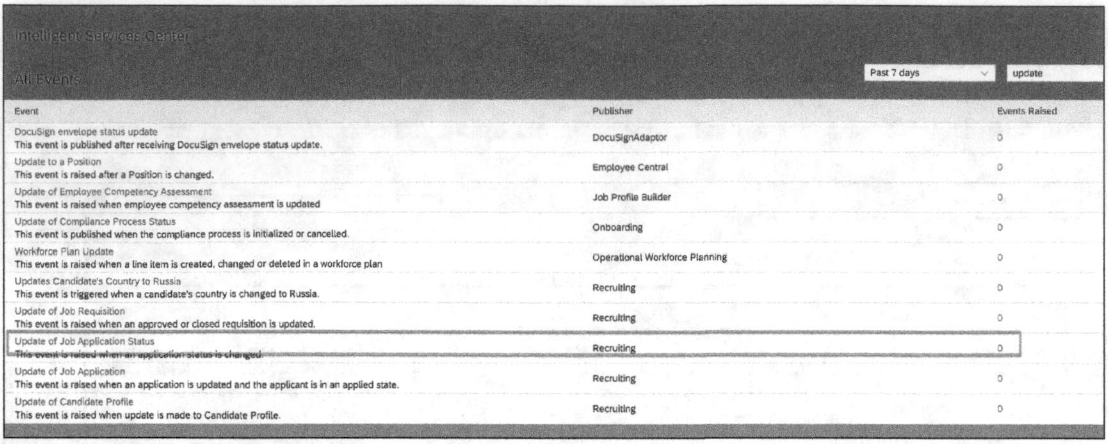

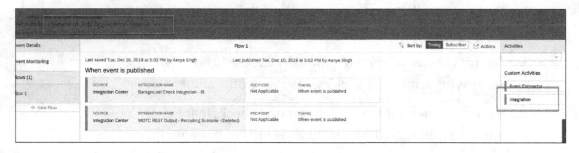

Event	Publisher	Events Raised
DocuSign envelope status update This event is published after receiving DocuSign envelope status update.	DocuSignAdaptor	0
Update to a Position This event is raised after a Position is changed.	Employee Central	0
Update of Employee Competency Assessment This event is raised when employee competency assessment is updated	Job Profile Builder	0
Update of Compliance Process Status This event is published when the compliance process is initialized or cancelled.	Onboarding	0
Workforce Plan Update This event is raised when a line item is created, changed or deleted in a workforce plan	Operational Workforce Planning	0
Updates Candidate's Country to Russia This event is triggered when a candidate's country is changed to Russia.	Recruiting	0
Update of Job Requisition This event is raised when an approved or closed requisition is updated.	Recruiting	0
Update of Job Application Status This event is raised when an application status is changed.	Recruiting	0
Update of Job Application This event is raised when an application is updated and the applicant is in an applied state.	Recruiting	0
Update of Candidate Profile This event is raised when update is made to Candidate Profile.	Recruiting	0

Figure 10-23. *Intelligent Services Center with Update of Job Application Status Highlighted*

Figure 10-24 shows where to create a new integration. In the beginning of this chapter, we mentioned that the underlying starting entity for the background check is JobApplication. You can associate an existing Integration Center template (ICD) or create a new one – you will be prompted by the system. You may refer to Chapter 7 on the Integration Center to create your integration or use an existing one.

Figure 10-24. *Screen Showing How to Add a New Integration to an Event*

Regardless of whether you create a new ICD or use an existing one, please be sure to click Save Flow as shown in Figure 10-25. If you don't save your flow, your background check will be initiated and the event will be raised, but your integration will not fire and consequently never reach the endpoint.

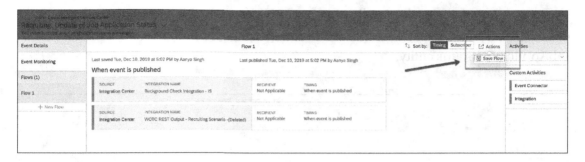

Figure 10-25. *Screenshot Indicating Where to Save Your Integration Flow*

Step 14: Setting Up the Integration Center Template (ICD) for Option 2

A couple of things to note when you set up the integration using Intelligent Services. You can always examine which integration template you have associated with the background check. Figure 10-26 shows how to identify which template you're using and launch your integration directly in the Integration Center.

Note   Clicking Add Rule will currently show an error. It's intended as a future functionality.

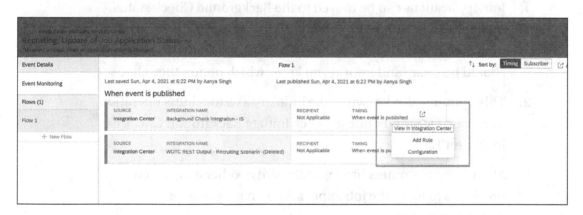

Figure 10-26. *Identify and Launch Directly into Integration Center*

In Step 3, we described how to modify your job requisition template. By definition, the Update of Job Application Status event will fire every time the status of the job application changes. It's therefore important to choose the correct filter. Figure 10-27 shows an example using the Application Status Name as the filter set to the value defined in step 3. You may also choose another property, appStatusId on the JobApplicationStatus entity.

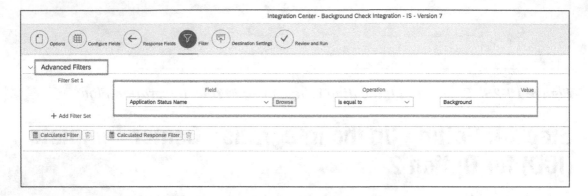

Figure 10-27. *Integration Center Template Showing the Filter on the Application Status Name*

Step 15: Moving Job Applications to Background Check Status

Notes:

1. Job Applications can be moved to the Background Check status manually from within the candidate summary for a bulk operation or moved individually within a job application. Your customer should be aware and inform you know which option they choose.

2. This option is also known in the industry as a touchless operation, because the recruiter doesn't click Initiate Background Check as in Option 1.

3. Many large customers have middleware or other automated processes to move the job applications en masse to the Background Check status.

4. There will be one event per update of job application status and the results can be viewed in the Intelligent Services Center in the following section.

Step 16: Intelligent Services Center – Checking the Results of Outbound Execution

If your integration is all set up, the next logical step is to change the status of the job application to Background Check and examine the results of your integration. The first step is to examine if an event associated with the status change has been raised. Figure 10-28 shows such an example showing that the Update of Job Application Status has indeed been raised and there are two possible events to examine. Click the number of executions in the right column.

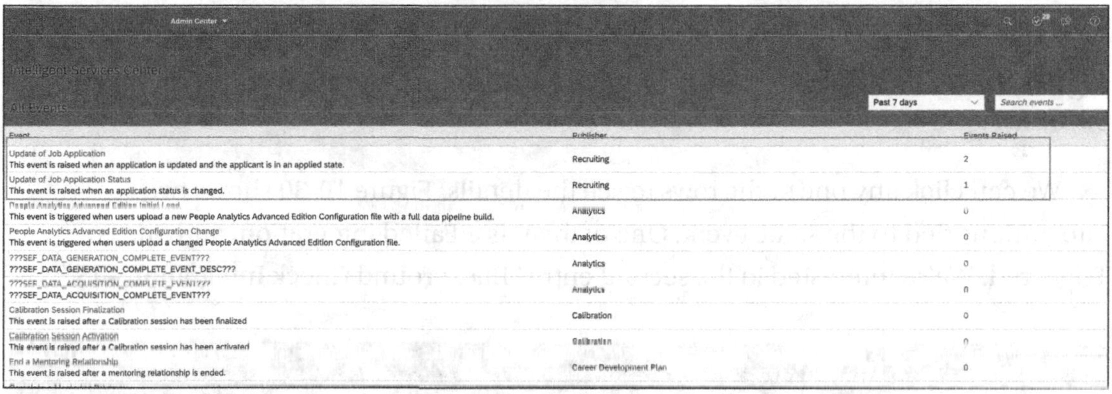

Figure 10-28. *Intelligent Services Center Showing Number of Events Raised Against Events*

The highlighted items show the two events for Update of Job Application Status Intelligent Services event. Click the number 1 to view the execution status as shown in Figure 10-29. There will be one event per update of job application status.

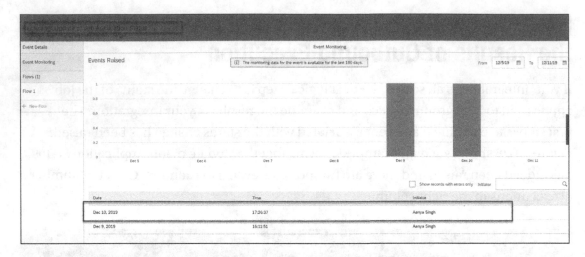

Figure 10-29. *Event Monitoring Showing a Sample Event Execution Summary*

We can click any one of the rows to see the details. Figure 10-30 shows two integrations tied to the same event. One of them is a Failed integration, and one is Triggered. We're interested in the second entry "Background Check Integration – IS."

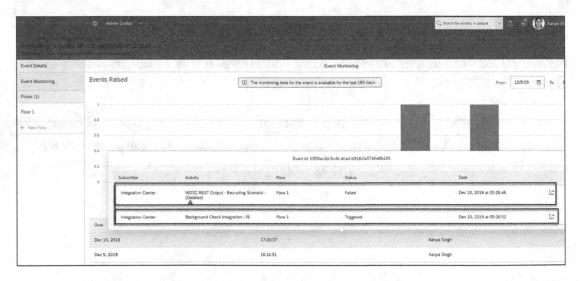

Figure 10-30. *Event Monitoring Showing Integration Center Executions per Event*

To view the results of the execution, click the ⬀ icon. This will directly launch the Execution Manager in the Integration Center. Figure 10-31 shows an example entry in the execution manager for the background check integration.

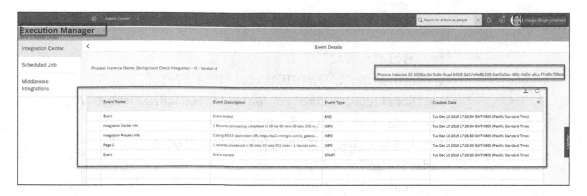

Figure 10-31. *Sample Illustration of the Execution Manager for the Raised Event for Background Check*

If you have gotten this far, your outbound integration for the background check has been set up, and it is now time to perform the API operations so that customers can receive the results from the background check vendor and view them in the recruiting user interface. Please refer to Chapter 1 on how to use and leverage ODATA APIs.

Step 17: Query JobApplicationBackground CheckRequest for Option 1

If the outbound Integration Center is set up successfully and the background vendor is able to receive the request and respond synchronously, the JobApplicationBackgroundCheckRequest should be completely populated:

```
GET: https://apisalesdemo4.successfactors.com:443/odata/v2/
JobApplicationBackgroundCheckRequest?$format=json&$filter=requestId eq 140
```

RESPONSE:

```
{
    "d": {
        "results": [
            {
                "__metadata": {
                    "uri": "https://apisalesdemo4.successfactors.com:443/
                    odata/v2/JobApplicationBackgroundCheckRequest(140L)",
```

```
                    "type": "SFOData.JobApplicationBackgroundCheckRequest"
                },
                "requestId": "140",
                "vendorOrderNo": "MYTHICAL_12112019_87654",
                "lastModifiedDateTime": "/Date(1576094482000+0000)/",
                "createdByUser": "sfadmin",
                "createdDateTime": "/Date(1575936725000+0000)/",
                "orderStatus": "OK",
                "applicationId": "3040",
                "responseDetail": "Order received and acknowledged.",
                "responseCode": "200",
                "vendorCode": "MYTHICAL",
                "jobApplication": {
                    "__deferred": {
                        "uri": "https://apisalesdemo4.successfactors.
                        com:443/odata/v2/JobApplicationBackgroundCheckReque
                        st(140L)/jobApplication"
                    }
                }
            }
        ]
    }
}
```

Notes:

1. requestId is the SuccessFactors-generated primary key for the
    JobApplicationBackgroundCheckRequest entity and cannot be
    changed. There will be one or more requests per background
    request.

2. vendorOrderId is the value set by the vendor as part of the
    synchronous response. You will use this vendorOrderId to send
    subsequent results.

3. vendorCode is set during the outbound Integration Center set up
    in Manage Data. Please refer to step 8 of this chapter.

4. orderStatus is sent as part of the response. The only two
   acceptable values are OK and ERROR. ERROR is generally set if
   the vendor is unable to respond to the synchronous request.

5. responseDetails and responseCode are also set by the vendor.
   These are for information purposes only.

Step 18: Creating the JobApplicationBackground CheckRequest Entity for Option 2

This step is applicable when the background check is initiated via Intelligent Services. Since the control is completely with the vendor, the record can be created via an API call. If there is no prior matching background check request, a new one is inserted and the response field editStatus shows INSERTED, else it's UPDATED.

The actual API call is: POST https://apisalesdemo4.successfactors.com:443/ odata/v2/upsert?$format=json

```
{
    "__metadata": {
        "uri": "JobApplicationBackgroundCheckRequest",
        "type": "SFOData.JobApplicationBackgroundCheckRequest"
    },
    "vendorCode": "MYTHICAL",
    "requestId": 140,
    "vendorOrderNo": "MYTHICAL_12112019_87654",
    "orderStatus": "OK",
    "applicationId": 3040,
    "responseCode": "200",
    "responseDetail": "Order received and acknowledged."
}
```

Response Payload:

```json
{
    "d": [
        {
            "key": "JobApplicationBackgroundCheckRequest/requestId=140",
            "status": "OK",
            "editStatus": "UPDATED",
            "message": "JobApplicationBackgroundCheckRequest Updated
            successfully for request 140",
            "index": 0,
            "httpCode": 204,
            "inlineResults": null
        }
    ]
}
```

The same notes apply to this operation as in step 17.

Step 19: Background Check – Retrieving Candidate Details

This section gives you an idea of how to construct your integration. The following sample uses the JobApplicationBackgroundCheckRequest as the top-level entity and navigates to other entities to retrieve additional information for an application. There's no guidance on whether to use the JobApplication or Candidate entities to construct your query. The choice is up to you. The response only shows the partial results.

```
GET: https://apisalesdemo4.successfactors.com:443/odata/v2/
JobApplicationBackgroundCheckRequest?$format=json&$select=vendorCode,
applicationId, jobApplication/firstName, jobApplication/country,
jobApplication/lastName, jobApplication/contactEmail, jobApplication/
dateOfBirth,jobApplication/address, jobApplication/address2,jobApplication/
city,jobApplication/gender,jobApplication/zip, jobApplication/
state,jobApplication/status,jobApplication/state/picklistLabels/
label&$filter=requestId eq '140'&$expand=jobApplication,jobApplication/
state/picklistLabels
```

Response:

```
{
    "d": {
        "results": [
            {
                "__metadata": {
                    "uri": "https://apisalesdemo4.successfactors.com:443/
                    odata/v2/JobApplicationBackgroundCheckRequest(140L)",
                    "type": "SFOData.JobApplicationBackgroundCheckRequest"
                },
                "applicationId": "3040",
                "vendorCode": "MYTHICAL",
                "jobApplication": {
                    "__metadata": {
                        "uri": "https://apisalesdemo4.successfactors.
                        com:443/odata/v2/JobApplication(3040L)",
                        "type": "SFOData.JobApplication"
                    },
                    "zip": "94123",
                    "firstName": "Theon",
                    "country": "United States",
                    "lastName": "Greyjoy",
                    "address": "100 Main Street",
                    "gender": "Male",
                    "contactEmail": "athanur+24@gmail.com",
                    "address2": null,
                    "city": "San Francisco",
                    "dateOfBirth": "/Date(534574800000+0000)/",
                    "status": "Open",
                    "state": {
                        "results": [
                            {
                                "__metadata": {
```

```
                                    "uri": "https://apisalesdemo4.
                                    successfactors.com:443/odata/v2/
                                    PicklistOption(11645L)",
                                    "type": "SFOData.PicklistOption"
                            },
                            "id": "11645",
                            "minValue": "-1",
                            "externalCode": "state_United_States_
                            California",
                            "maxValue": "-1",
                            "optionValue": "-1",
                            "sortOrder": 20,
                            "mdfExternalCode": "state_United_States_
                            California",
                            "status": "ACTIVE",
                            "parentPicklistOption": {
                                "__deferred": {
                                    "uri": "https://apisalesdemo4.
                                    successfactors.com:443/odata/
                                    v2/PicklistOption(11645L)/
                                    parentPicklistOption"
                                }
                            },
                            "picklistLabels": {
                                "results": [
                                    {
                                        "__metadata": {
                                            "uri": "https://
                                            apisalesdemo4.
                                            successfactors.com:443/
                                            odata/v2/PicklistLabel
                                            (locale='en_US',
                                            optionId=11645L)",
                                            "type": "SFOData.
                                            PicklistLabel"
```

```json
        },
        "optionId": "11645",
        "locale": "en_US",
        "id": "12076",
        "label": "California",
        "picklistOption": {
            "__deferred": {
                "uri": "https://
                apisalesdemo4.
                successfactors.
                com:443/odata/v2/
                PicklistLabel
                (locale='en_US',
                optionId=11645L)
                /picklistOption"
            }
        }
    },
    {
        "__metadata": {
            "uri": "https://
            apisalesdemo4.
            successfactors.com:
            443/odata/v2/
            PicklistLabel
            (locale='fr_FR',
            optionId=11645L)",
            "type": "SFOData.
            PicklistLabel"
        },
        "optionId": "11645",
        "locale": "fr_FR",
        "id": "120541",
        "label": "Californie",
```

```
                                    "picklistOption": {
                                        "__deferred": {
                                            "uri": "https://
                                            apisalesdemo4.
                                            successfactors.com:
                                            443/odata/v2/
                                            PicklistLabel
                                            (locale='fr_FR',
                                            optionId=11645L)
                                            /picklistOption"
                                        }
                                    }
                                },
                    }
(...only partial response shown...)
        ]
    }
}
```

Step 20: Inserting Records into JobApplicationBackgroundCheckResult

Note: SuccessFactors does not imply a specific structure in how the package is constructed since the solution has to apply to all vendors. The tighter the specification, the more difficult it will be for a third-party vendor to implement. In this example, we're going to mock up a background check request, and it is completely arbitrary.

The structure is as follows:

Comprehensive Background (main package) – the API call related to this will contain a dummy reportUrl that shows up in the UI:

- Driver's License (sub-package)

- Employment (sub-package)

There is no implied hierarchy of packages. So, we have set up a pseudo parent-child relationship. You may use this as an example or simply use one package per background check.

POST: https://apisalesdemo4.successfactors.com:443/odata/v2/JobApplicationBackgroundCheckResult?$format=json

Request Payload:

```json
{

    "vendorOrderNo": "MYTHICAL_12112019_87654",
    "vendorCode": "MYTHICAL",
    "stepName": "Driver's License",
    "stepStatus": "Received",
    "stepMessage": "Background Check is in process. Please check back
                    later",
    "finalStep": false

}
```

Response Payload:

```json
{

    "d": {
        "__metadata": {
            "uri": "https://apisalesdemo4.successfactors.com:443/odata/v2/
            JobApplicationBackgroundCheckResult(140L)",
            "type": "SFOData.JobApplicationBackgroundCheckResult"
        },
        "vendorOrderNo": "MYTHICAL_12112019_87654",
        "vendorCode": "MYTHICAL",
        "stepName": "Driver's License",
        "stepStatus": "Received",
        "stepMessage": "Background Check is in process. Please check back
                        later",
        "finalStep": false,
        "statusId": "140"
    }

}
```

Notes:

1. stepName and stepStatus are treated as name-value pairs and are visible in the UI. It's entirely up to the vendor to define how it's constructed.

2. The system will only check if the stepName exists in a previous call. Pay close attention to this value. "Driver's License" is not the same as "Drivers License." The system will create two entries in the UI, one for "Driver's License" and another for "Drivers License."

3. stepMessage is only visible in the ODATA audit logs and it's for informational purposes only.

4. finalStep is intended to show that the result is complete, although there's no technical limit on how many times you set the value to true or false.

5. statusId is a system-generated value and cannot be updated.

Figure 10-32 shows a sample UI screen to confirm the results of an API call.

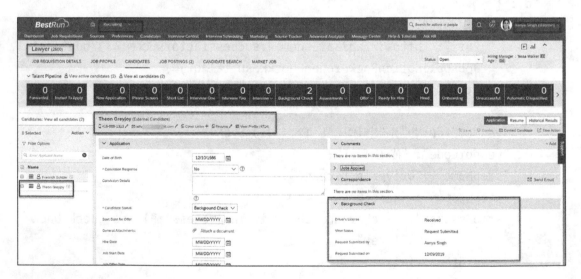

Figure 10-32. *UI Confirmation of the API Call*

The following is an example of updating the results of the prior call:

POST: https://apisalesdemo4.successfactors.com:443/odata/v2/JobApplicati onBackgroundCheckResult?$format=json

Request Payload:

```json
{

    "vendorOrderNo": "MYTHICAL_12112019_87654",
    "vendorCode": "MYTHICAL",
    "stepName": "Driver's License",
    "stepStatus": "Completed - no discrepancies",
    "stepMessage": "Background Check is in process. Please check back
                    later",
    "finalStep": false
}
```

Response:

```json
{
    "d": {
        "__metadata": {
            "uri": "https://apisalesdemo4.successfactors.com:443/odata/v2/
            JobApplicationBackgroundCheckResult(141L)",
            "type": "SFOData.JobApplicationBackgroundCheckResult"
        },
        "vendorOrderNo": "MYTHICAL_12112019_87654",
        "vendorCode": "MYTHICAL",
        "stepName": "Driver's License",
        "stepStatus": "Completed - no discrepancies",
        "stepMessage": "Background Check is in process. Please check back
                        later",
        "finalStep": false,
        "statusId": "141"
    }
}
```

Figure 10-33 shows the UI confirmation of the results of the second API call.

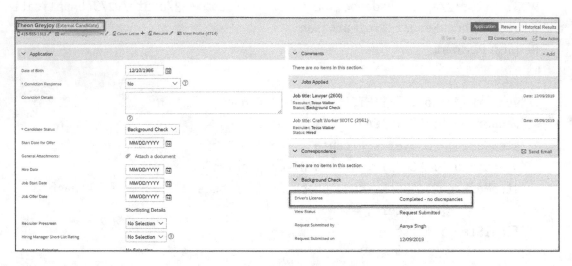

Figure 10-33. *UI Confirmation of the Results of the API Call (Second Screenshot)*

The following API call builds on the arbitrary package structure defined in the beginning of this step:

POST: https://apisalesdemo4.successfactors.com:443/odata/v2/JobApplicationBackgroundCheckResult?$format=json

Request Payload:

```
{
    "vendorOrderNo": "MYTHICAL_12112019_87654",
    "vendorCode": "MYTHICAL",
    "stepName": "Employment",
    "stepStatus": "Submitted - pending verification",
    "stepMessage": "Background Check is in process. Please check back
                later",
    "finalStep": false
}
```

Response:

```
{
    "d": {
        "__metadata": {
            "uri": "https://apisalesdemo4.successfactors.com:443/odata/v2/
            JobApplicationBackgroundCheckResult(142L)",
            "type": "SFOData.JobApplicationBackgroundCheckResult"
        },
        "vendorOrderNo": "MYTHICAL_12112019_87654",
        "vendorCode": "MYTHICAL",
        "stepName": "Employment",
        "stepStatus": "Submitted - pending verification",
        "stepMessage": "Background Check is in process. Please check back
                        later",
        "finalStep": false,
        "statusId": "142"
    }
}
```

Figure 10-34 shows the UI confirmation of the stepName called Employment.

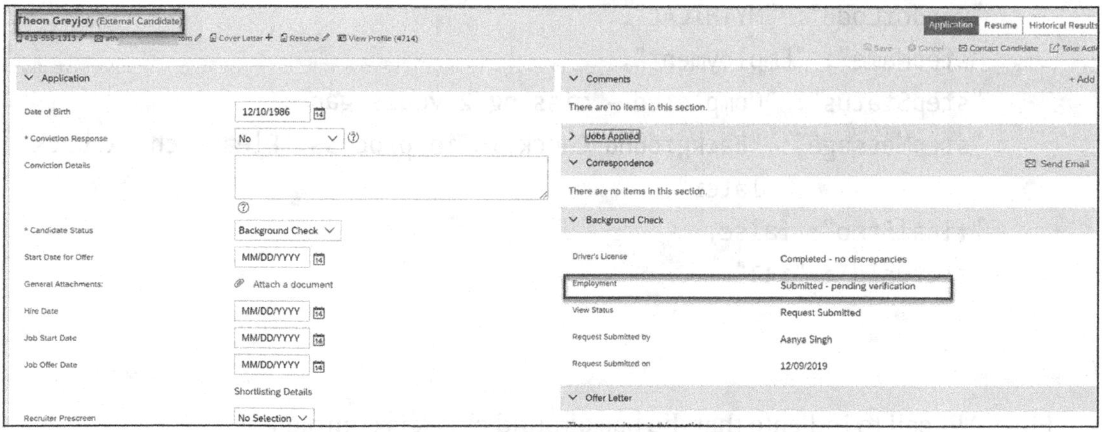

Figure 10-34. *UI Confirmation for Second "Package" Sent.*

The following simply updates the previous API call:

POST: https://apisalesdemo4.successfactors.com:443/odata/v2/JobApplicationBackgroundCheckResult?$format=json

Request Payload:

```
{
    "vendorOrderNo": "MYTHICAL_12112019_87654",
    "vendorCode": "MYTHICAL",
    "stepName": "Employment",
    "stepStatus": "Completed - missing 2 years gap",
    "stepMessage": "Background Check is in process. Please check back
                    later",
    "finalStep": false
}
```

Response:

```
{
    "d": {
        "__metadata": {
            "uri": "https://apisalesdemo4.successfactors.com:443/odata/v2/
            JobApplicationBackgroundCheckResult(143L)",
            "type": "SFOData.JobApplicationBackgroundCheckResult"
        },
        "vendorOrderNo": "MYTHICAL_12112019_87654",
        "vendorCode": "MYTHICAL",
        "stepName": "Employment",
        "stepStatus": "Completed - missing 2 years gap",
        "stepMessage": "Background Check is in process. Please check back
                        later",
        "finalStep": false,
        "statusId": "143"
    }
}
```

Final API call to indicate that the background check is complete:

POST: https://apisalesdemo4.successfactors.com:443/odata/v2/JobApplicationBackgroundCheckResult?$format=json

Request Payload:

```
{

    "vendorOrderNo": "MYTHICAL_12112019_87654",
    "vendorCode": "MYTHICAL",
    "stepName": "Comprehensive Background",
    "stepStatus": "Completed - Missing information. Proceed with
                  caution.",
    "stepMessage": "Background Check completed.",
    "reportUrl": "https://partneredge.sap.com/en/partnership/manage/
                 mmp.html",
    "finalStep": true
}
```

Response Payload:

```
{
    "d": {
        "__metadata": {
            "uri": "https://apisalesdemo4.successfactors.com:443/odata/v2/
            JobApplicationBackgroundCheckResult(144L)",
            "type": "SFOData.JobApplicationBackgroundCheckResult"
        },
        "vendorOrderNo": "MYTHICAL_12112019_87654",
        "vendorCode": "MYTHICAL",
        "stepName": "Comprehensive Background",
        "stepStatus": "Completed - Missing information. Proceed with
                      caution.",
        "stepMessage": "Background Check completed.",
        "reportUrl": "https://partneredge.sap.com/en/partnership/manage/
                     mmp.html",
        "finalStep": true,
        "statusId": "144"
    }
}
```

Figure 10-35 shows the final result of the ODATA API call indicating that the background check is complete.

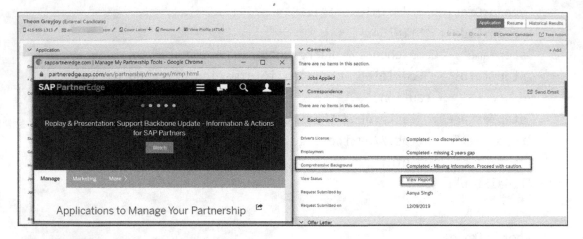

Figure 10-35.   *Screenshot Showing the Completed Background Check with the Results URL*

Conclusion

If you have followed the steps listed earlier, you should be in a position to successfully build a background check integration for an SAP SuccessFactors Recruiting customer in two different ways. The first is when a recruiter or another authorized user initiates the action within the Job Application User Interface. The second is when a job application is moved to the Background Check status using a combination of Intelligent Services and Integration Center.

FAQs

1. Can two or more background check vendors coexist in an SAP SuccessFactors Recruiting instance?

2. Yes, but with a caveat. Only one of the background check vendors can leverage the background check portlet. In other words, the results passed by only one vendor will be displayed on the portlet. The second vendor has to add custom fields on the job application to return results.

3. Can you build a background check integration without Integration Center?

4. If you're going to use the background check initiated from within the job application UI, Integration Center must be used. There is no limitation if Intelligent Services are used.

5. Can I launch a background check at different steps in the candidate pipeline?

6. Yes, you can, as long as the feature permission in step 3 is repeated for each status in the applicant workflow.

7. What happens when a customer has more than one job requisition template or job application template?

8. The Integration Center templates have to accommodate for each applicable status for each template.

9. What if the background check vendor cannot support synchronous JSON request/response?

10. The Integration Center templates will not have response fields mapping. In that case, the vendor has complete control over how the JobApplicationBackgroundCheckRequest object is constructed.

11. What are the authentication mechanisms supported by Integration Center?

12. Integration Center supports the following authentication mechanisms, none, Basic Authentication or OAuth 2.0 with SAML Flow, OAuth 2.0 with Grant Type as Password, and OAuth2.0 with Grant Type as Client Credentials. Please note that Basic Authentication will be deprecated in November 2021 and fully decommissioned by December 2022.

13. How do I troubleshoot the integration?

14. If the integration is launched from the job application UI via Initiate Background Check, the starting point will be the Execution Monitor in Integration Center. There will be one execution per background check. The Execution Monitor will provide a sequential list of entries to identify the source of the problem. If the integration uses Intelligent Services, the starting point should be Intelligent Services Center ➤ Update of Job Application Status Event ➤ Event Monitor, and optionally the Execution Monitor of Integration Center (if that is employed as part of integration).

CHAPTER 11

Candidate Offer Automation Using Business Rules

As we have progressed throughout the book, we've shown you a variety of integration and automation methods at each stage of the recruiting process. Thus far, each chapter of this book has focused on an integration or automation that involves a third-party system, creating a file, or making/receiving an external call. However, there are some automations that can be executed entirely within the SuccessFactors system. As we approach the offer stage of the recruiting process, we take this opportunity to review such automations since creating an offer is less likely to require integration (outside of standard integrations such as document signature). A prime example of these types of internal automations is business rules. Let's take a look!

Business Rules Overview

Business rules are a feature of SAP SuccessFactors that allow administrators to construct If/Then conditional logic and then execute a resulting scripted behavior when the conditions are met. This feature is not available for all objects and data in the system, but the list is expanding with every release. In general, business rules for scenarios where fields are initialized, changed, and saved will work on job requisitions, offers, candidates, and applications. However, there are a lot of exceptions we have covered in the following note.

© Anand 'Andy' Athanur, Mark Ingram and Michael A. Wellens 2022
A. A. Athanur et al., *Innovative SAP SuccessFactors Recruiting*, https://doi.org/10.1007/978-1-4842-7425-5_11

Note *For full details on managing business rules in SAP SuccessFactors Recruiting, see SAP notes 2618607 and 2515173. For details on supported field types for business rules for job requisitions, see SAP note 2675870. For details on business rules for offers, see SAP note 2860149. For details on business rule scenarios on applications that will not trigger, see SAP note 2866437. For details on business rule scenarios on candidates that will not trigger, see SAP note 2781775.*

In general, we do not recommend business rules geared toward candidates making updates because these are not compatible with the mobile apply feature. In most scenarios, the advantages allowing candidates the ease of applying on their mobile device outweigh the advantages of a business rule on these objects.

Typically, the result of a business rule when conditions are met is to update the value of a field or execute another process or rule. Written in English, an example business rule would be: "When the offer details screen loads, then copy the value of field X to field Y." Another example would be: "When the job requisition is updated, if field X is greater than field Y, then copy field X to field Z; otherwise, copy field Y to field Z." We will get into the specifics of how a statement such as the ones mentioned earlier is converted into a business rule configuration later in this chapter.

Configuring Business Rules

Thus far, we've given you a brief explanation of business rules and a multitude of SAP notes to review. It can be daunting to understand how business rules work and what their limitations are. The best way to fully understand is to dive into the system and take a look for ourselves. Let's get started on a quick walk-through!

Our very simple scenario will warn the user when they are making changes if the requisition is more than five weeks old. To get started, we will first create a simple warning message for users:

1. Type and select "Manage Data" in the search bar.

2. Click the drop-down.

3. Fill in the screen as shown in Figure 11-1. For the "Text" field, type "Warning, this requisition is already late! Please avoid significant changes!" Set "External Code" to "LATE_REQUISITION" and set "Name" to "Late Requisition." Make sure the "Status" is set to "Active."

4. Click "Save."

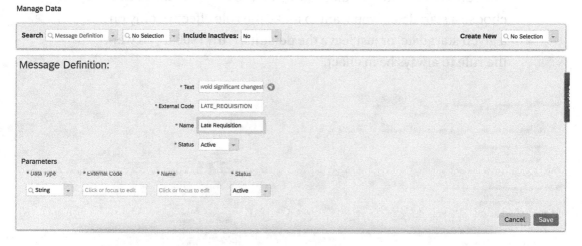

Figure 11-1. Example Message Definition

Next, we will make our business rule. To access business rules for recruiting, follow these steps:

Note   *Be sure you have permissions and provisioning is set up. Please reference SAP note 2515173 for prerequisites.*

---

1. Type and select "Configure Business Rules" in the search bar. The Business Rules Admin screen will load.

2. Click the "+" icon in the upper-right-hand corner. The Create Business Rule screen will load.

3. You will see a list of different business rule types on the left. Recruiting rules are limited to the "Basic" type seen down at the bottom for Offer and Requisition objects. Click the arrow next to "Basic" and then click the radio button next to "Basic."

Note   *You can also choose "Configure Business Rules on Candidate Profile" or "Job Application: Basic Business Rule Scenario" under "Recruiting" if you want to make a business rule for either of those objects.*

4. The screen will update as seen in Figure 11-2. Type in a rule name. The rule ID will auto-populate based on the rule name. You can choose a start date if you want to make the rule effective only on a given start date, or just leave the default 01/01/1900 if you want the rule to always be in effect.

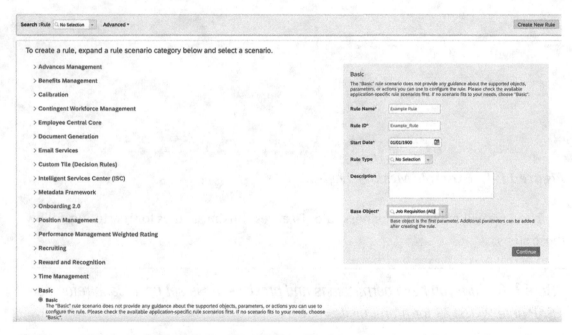

Figure 11-2.   *Creating an Example Business Rule*

5. Choose a base object. In our example, we choose "Job Requisition (All)" which refers to any job requisition template (rather than individual templates also listed).

6. Click "Continue." The rule details screen will load as seen in Figure 11-3.

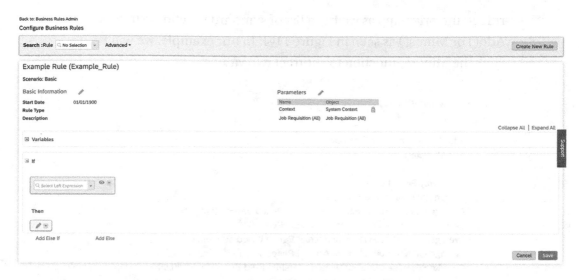

Figure 11-3. *Business Rule Details Screen*

7. At the top of the screen, you can click the pencil icon next to "Basic Information" to edit the metadata you set up in the prior screen. You can also click the pencil icon next to "Parameters" to bring in information from other areas of the system (e.g., other objects). By default, you will have system context (if you are familiar with JAVA or similar programming, the context is essentially who the logged on user is from which we can also look up other information about the user) and the base object you choose in the last screen as chosen parameters (in our case the Job Requisition). For the sake of this walk-through, we will leave these alone so that the Job Requisition and Context are brought in as parameters.

8. You can click the "+" icon next to "Variables" to expand the variables area. Much like in JAVA, C++, or other object-oriented programming, this area allows you to create variables that can contain values which can be used in your If/Then logic of the business rule. The advantage of using a variable is that it will be available across the scope of all of your If and If/Then expressions/conditions. Give the variable a name by typing one into the open field. Then click the field to the right of the "=" sign. You will see the list of parameters you brought into the business

285

rule in the prior step as well as a list of standard functions such as Add() or Minus() as seen in Figure 11-4. In our example, we will click the Minus() function to subtract two values.

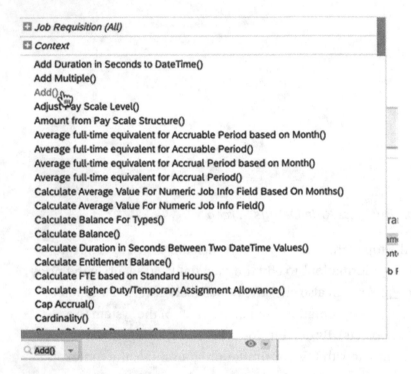

Figure 11-4. *List of Parameters and Standard Functions*

9. An example is shown in Figure 11-5 where we have created a variable called var_FiveWeeksAgo that uses the Minus() standard function to calculate what week of the month it was five weeks ago. Fill in the screen as shown in Figure 11-5. While this variable would probably be constant on the day of execution, we could also use variables for things like calculating an ongoing total throughout the logic.

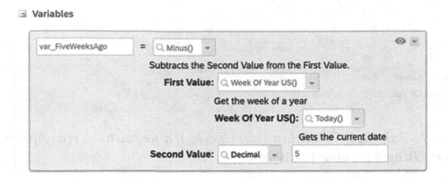

Figure 11-5. *Example Variable*

10. The following variables section is where you can build out the "If" area. This is the conditional logic that determines whether the "Then" logic is executed or not. In Figure 11-6, we can see there is an option to set this to "Always True" so that the "Then" logic is always executed.

Figure 11-6. *Example Setting "If" to "Always True" So That "Then" Logic Will Always Execute*

11. If we do not want the logic to always execute, then we must come up with some conditional logic. Like we did creating a variable, clicking the field next to "If" will show the list of objects you have brought into the business rule as parameters (in our case the Requisition Object as well as the Context) as well as a list of standard functions similar to what we saw in Figure 11-4. For our walk-through, we will use our variable to compare if the Job Requisition was approved five or more weeks ago. Fill in the If logic as shown in Figure 11-7.

Figure 11-7. *Sample If Condition That Checks If a Requisition Was Approved Five Weeks Ago Using a Sample Variable*

12. Next, we define what will happen when the condition is met. When you click the first box, you can see there are a variety of functions we can execute. We can set values of fields within the parameters we selected, we can raise a message, we can create or delete objects, we can execute other functions, or we can add to other objects. This is illustrated in Figure 11-8. In our walk-through, we will raise the warning message we created earlier in this section. In the first box, choose "Raise Message."

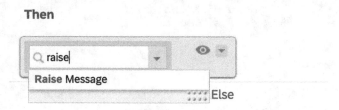

Figure 11-8. *Available Actions to Initiate Within "Then" Section*

13. Next, we will choose the message we created. In the "Message" field, choose "LATE_REQUISITION" and set "Severity" to "Warning." An example is shown in Figure 11-9.

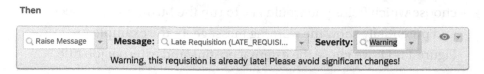

Figure 11-9. *Example "Then" Logic*

14.   Click "Save."

Congratulations! You have completed defining your business rule! The final rule logic should look as shown in Figure 11-10.

Figure 11-10. *Final Overall Business Rule Logic*

As a final step, we now have to associate our business rule with a requisition object and define when it will execute. Follow these steps:

1.   Type and select "Manage Rules in Recruiting" in the search bar.

2.   The screen should default to the Job Requisition object. Choose the specific Job Requisition template on which you would like to have the rule active.

3.   Business rules can be executed whenever a field changes; the template is first loaded on screen, or whenever it is saved. We will want this warning to show up any time the requisition is changed. Therefore, under the "Field Change Rules" section,

choose which fields you would like to run the business rule check and potentially show the warning message. In the example in Figure 11-11, we chose the external posting and external job title.

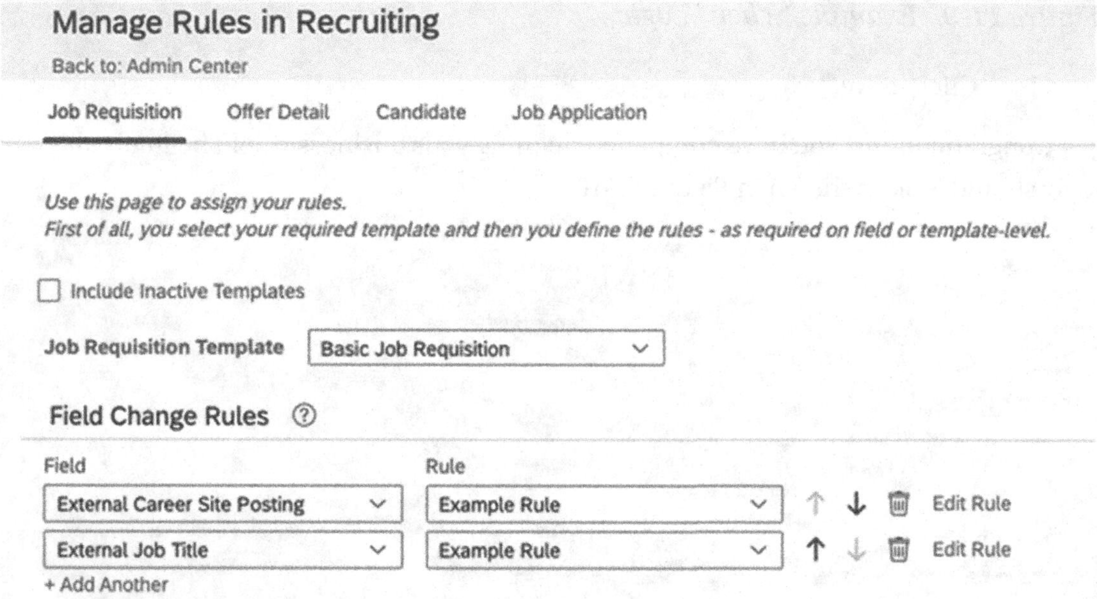

Manage Rules in Recruiting

Back to: Admin Center

Job Requisition   Offer Detail   Candidate   Job Application

Use this page to assign your rules.
First of all, you select your required template and then you define the rules - as required on field or template-level.

☐ Include Inactive Templates

Job Requisition Template   [Basic Job Requisition   ⌄]

Field Change Rules ⑦

Field	Rule				
[External Career Site Posting ⌄]	[Example Rule ⌄]	↑	↓	🗑	Edit Rule
[External Job Title ⌄]	[Example Rule ⌄]	↑	↓	🗑	Edit Rule

+ Add Another

Figure 11-11. *Associating a Business Rule with a Field Change on a Particular Requisition Template*

4. Click "Save."

Great job! You have now set up your first business rule! We should now be able to test out changing the fields you specified and immediately getting the warning message you set up. An example is shown in Figure 11-12 where we have changed the "External Job Title" field on an existing requisition that was approved five or more weeks ago and received the warning message as expected.

Figure 11-12. *Wanting Message Received After Changing "External Job Title" Field on Existing Requisition Meeting Rule Criteria*

Relevant Business Scenarios

Now that we understand how business rules work, we can see there are countless relevant business scenarios where they can provide a practical solution. Let's take a closer look at a few.

In our walk-through, we raised a warning message in the event that edits were being made to an older requisition. In a similar scenario, we could raise warning messages if a salary field was set out of bounds. We might also raise a warning if the number of openings is set too high and needs additional authorization. Additionally, we could also cause a hard stop on save and let the user know that the salary is restricted or that the number of openings past a certain point requires more than one requisition to be approved.

In addition to simple warning messages or hard stops, we can also calculate new values or alter other fields based on the value of a prior field. For example, if a user selects a certain Company Code or Division and those only have one location, we could automatically set the location field to that one and only one location.

We can also take this to the next level and have data pass from system to system and object to object to have an effect on a downstream process. In this comprehensive example scenario, an HRIS system creates a requisition and supplies it with values for both a minimum and maximum salary field. In addition, a candidate also receives an assessment score as part of the recruiting process prior to extending an offer. When a recruiter opens up an offer, we can use business rules to factor in the minimum salary, maximum salary, and the assessment score to come up with a recommended offer. We will step through this in greater detail in the next section as we set up a business rule to make this happen.

Case Study: Calculating a Recommended Offer Amount

Now that we've seen what business rules can do and have a good comprehensive business scenario, we are ready to start building it! First, we will take a look at our business scenario before teaching you how to build the solution.

Business Scenario

To quickly review, in our scenario, an HRIS system has passed minimum and maximum recommended salary values to our requisition as part of a requisition creation interface like you have built in Chapter 2. In addition, an assessment vendor has passed an assessment score to our application such as what has been built in Chapter 8. Using these as inputs, we can now build a rule that populates a recommended salary on the offer template. A diagram of this data flow is shown in Figure 11-13.

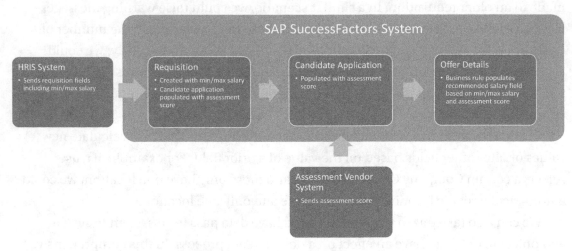

Figure 11-13. *Data Flow of Salary Recommendation Case Study*

Implementing a solution such as the one shown in Figure 11-2 would save recruiters a lot of time, ensure the data is accurate and not prone to human error, and increase moral by not having to deal with these headaches. Without such a solution, when recruiters go to the offer details stage, they would have to log into the HRIS system to look up min/max salary, then log into the assessment vendor system, and look up the candidate's scores. Then, they would likely have to paste the data in a spreadsheet and use a calculation to come up with final salary recommendation. Then they would have to paste this recommendation in SuccessFactors. That is a lot of steps and places where errors could be made! Instead, in our solution, the recommendation is made for the recruiter automatically. This will help save recruiters significant time and frustration!

Implementing the Solution

To get started creating the business rule, follow the given steps:

Note   *In this example, we assume that integrations have already been built to populate the "Salary Min" and "Salary Max" fields on the requisition and the "Average Rating" field on the Job Application.*

---

1. Type and select "Configure Business Rules" in the search bar. The Business Rules Admin screen will load.

2. Click the "+" icon in the upper-right-hand corner. The Create Business Rule screen will load.

3. You will see a list of different business rule types on the left. Remember that recruiting rules are limited to the "Basic" type seen down at the bottom for Offer and Requisition objects. Click the arrow next to "Basic" and then click the radio button next to "Basic."

4. Enter a rule name and use the default rule ID and start date. Enter the base object as "Offer Approval (All)." Click "Continue." An example is shown in Figure 11-14.

Figure 11-14. *Example Metadata for Offer Approval Business Rule*

5. We can leave the parameters as default and begin to focus on our variables. We will first need to set up the calculations of recommended salary levels. We will say that a lower salary is set at 1.1 times the minimum, a middle-level salary is 1.25 times the minimum, and a higher-level salary is .9 times the maximum. Start by naming each variable, then choosing the multiply function, and then choosing the minimum or maximum salary field to multiply by the appropriate number. When you click the box next to the "Value" field, you will see the Job Requisition object shows up as an available object beneath the Offer Approval object so that you can select the Salary Min or Salary Max fields within it. An example of the fine set of variables is shown in Figure 11-15.

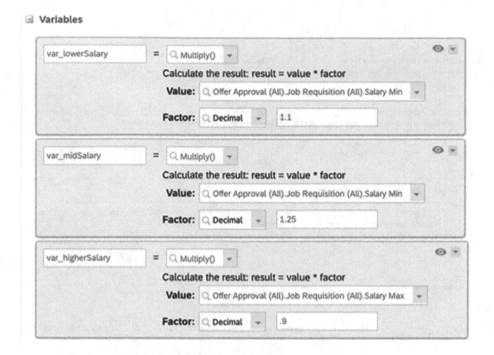

Figure 11-15.   *Example Variables Declaring Salary Levels*

6.   Now that our variables are set up, we can build our If/Then/Else
     logic. We will assume the assessment rating scale is a five-point
     scale. We will want to declare that if the rating on the application
     is lower than 1, we will set the recommended base salary to be the
     minimum salary from the requisition. If the rating is 1 or higher
     but lower than 2, we will recommend the lower salary variable.
     If the rating is greater than or equal to a 4, then the higher salary
     variable is set for the recommended salary. In all other cases, the
     salary will be set to the middle salary recommendation. To do this,
     build your logic as shown in Figure 11-16 as we did in our business
     rules walk-through. (Hint: To add the "and" to your If statement,
     click the down carrot icon on the far right of the expression and
     choose "Add" and then "Parent And" as shown in Figure 11-17.)

Figure 11-16. *Example If/Then/Else Logic Setting Recommended Base Salary*

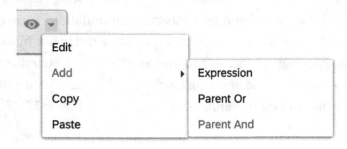

Figure 11-17. *Adding a "Parent And"*

7. Be sure to click "Save" to save your work.

8. Now we will associate our business rule with our available offer templates and make sure it executes when the template loads so that recruiters can see the recommended salary on the form as they fill it out. Type and select "Manage Rules in Recruiting" in the search bar.

9. When the "Manage Rules in Recruiting" screen appears, click the "Offer Detail" tab.

10. Under "Template-level Rules," click "Add Another" and then choose your business rule. Your final result should look like the example in Figure 11-18.

Figure 11-18. *Associating the Offer Amount Rule to an Offer Template to Load Upon Initialization*

11. Click "Save."

Great job! You've got your business rule set up, and now you can test it out! When you select a requisition that has the "Salary Min" and "Salary Max" fields populated (and optionally an assessment score since by default we will recommend a mid-level salary if there is no score), we can see that the "Base Salary" field auto-populates with our calculated recommendation as soon as we create an offer detail. An example is shown in Figure 11-19.

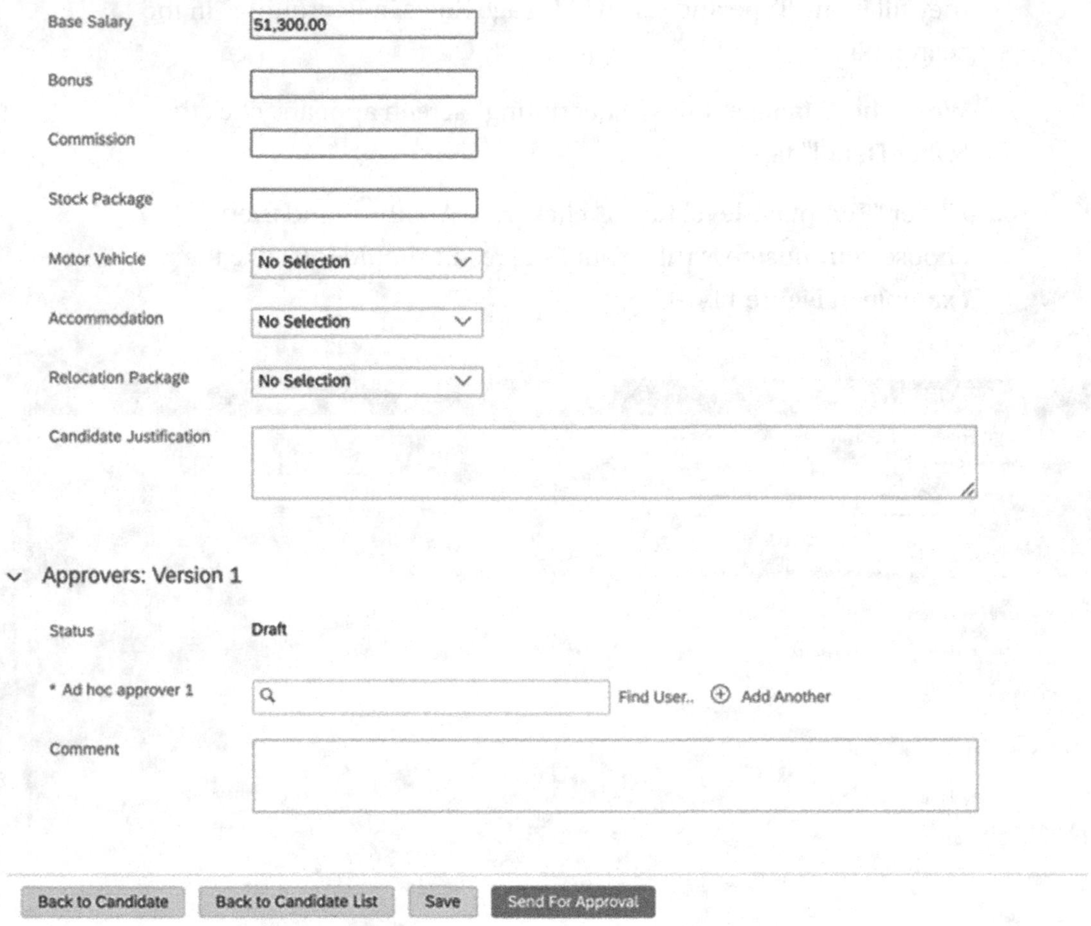

Figure 11-19. *Example New Offer Approval Form with "Base Salary" Field Auto-populated Based on Business Rule Logic*

Congratulations! With your new business rule in place, recruiters will no longer have to calculate the recommended salary offer on their own. You have now set up a time-saving business rule building on the knowledge you have obtained across multiple chapters! You can now see how business rules can be used independently or in conjunction with other integrations to provide useful automations.

Conclusion

We hope you have enjoyed this chapter on how business rules can help with a variety of recruiting processes – in particular automating the offer process. We started out introducing the concept of business rules and how they can help automate certain functions within SuccessFactors. We then stepped into a walk-through of creating a very simple business rule while also pointing out how the different features of business rules work. Hopefully, this got your technical creativity flowing as we reviewed some plausible business scenarios where this functionality would be useful. We then stepped through a comprehensive case study that showed how business rules could be used to manipulate data linked across systems and objects to arrive at a recommended job offer automation. Indeed, this real-life scenario would save recruiters a lot of time and frustration logging on to multiple systems and performing manual calculations. Instead, we taught you how to configure SAP SuccessFactors to make life easier for your recruiters. We hope you found it rewarding to see how linking the features and system processes introduced across the chapters can be used to build a very useful end result in a downstream process!

CHAPTER 12

Intelligent Services

Enterprise applications have been in use for a long time, and as time has progressed, the need to integrate has become central to any enterprise application. In the 1990s, the primary mechanism for extracting data and adding data was FTP and its recent variant Secure FTP. With the advent of Java as a programming language in the mid-1990s, messaging has been part and parcel of the programming language, primarily Java Message Queue. Over time, this methodology has spread to several enterprise applications including SuccessFactors.

This term is commonly referred to in the industry as "webhooks." Wikipedia defines webhooks as *"user-defined HTTP callbacks. They are usually triggered by some event, such as pushing code to a repository or a comment being posted to a blog. When that event occurs, the source site makes an HTTP request to the URL configured for the webhook. Users can configure them to cause events on one site to invoke behavior on another."* So, what exactly is Intelligent Services? Intelligent Services simplifies the user experience for employees, managers, and business partners by integrating into a single experience the individual transactions in a multistep HR process that cross the traditional boundaries of HCM applications and organizational departments. Intelligent Services links with external systems that allow seamless integration with third-party applications. For example, if a hiring manager goes on leave, not only does the core HR system require updates, but potentially other systems such as the Applicant Tracking System, performance management system, learning system, payroll system, benefits carrier, and more.

End users often cannot navigate the complexity of Human Experience Management (HXM), and companies are forced to spend large sums of money for shared HR services or business process outsourcing to ensure that all downstream processes are completed. With Intelligent Services, users no longer figure out what's next, or which system to access, or rely on shared HR services or business partners. Less help is required from shared HR services, reducing the workload and cost of HR shared services.

© Anand 'Andy' Athanur, Mark Ingram and Michael A. Wellens 2022
A. A. Athanur et al., *Innovative SAP SuccessFactors Recruiting*, https://doi.org/10.1007/978-1-4842-7425-5_12

At the heart of Intelligent Services are events, publishers, and subscribers. Different modules in SAP SuccessFactors raise events, for example, when a new hire is completed, a job requisition is updated, or when a learning event is completed. Each event has a publisher, the underlying SAP SuccessFactors module. An external system (or even another SAP SuccessFactors module) can subscribe to an event and react to it to perform downstream activities. Each publisher defines the event structure and the associated payload.

Step 1: Permissions in Admin Tools – Enable Intelligent Services

Log into the customer's instance as an administrator and navigate to Admin Tools ➤ Manage Permission Roles. Choose System Admin as the Permission Role and ensure the following are selected. The SuccessFactors Recruiting system requires several permissions to be set up for different roles. Table 12-1 shows the applicable list of permissions.

Table 12-1. *Necessary Role Based Permissions for Intelligent Services*

System Role	Permission Category	List of Permissions
API User	General User Permissions	– User Login – SFAPI User Login
API User	Intelligent Service Tools	– Intelligent Services Center (ISC)
API User	Manage Integration Tools	– Access to Event Notification Subscription – Access to Event Notification Audit Log – Access to ODATA API Audit Log – Access to API Center – Access to ODATA API Metadata Refresh and Export – Access to ODATA API Data Dictionary – Access to Integration Center – Access to Outbound Trust Manager
API User	Metadata Framework	– Admin Access to MDF ODATA API

If the customer instance you're working in was created before Q1 2017 release, you may have to explicitly enable Intelligent Services. If this is the case, you have to navigate to Admin Center ➤ Upgrade Center. Search for Recommended Upgrades and locate Intelligent Services and choose Learn More and Upgrade Now. Confirm your changes and Intelligent Services should be enabled.

If the customer instance was created after Q1 2017, Intelligent Services was enabled by default, but you need to ensure that the business rules were imported. This step applies for Employee Central, but it is a good idea to do it, nevertheless. Navigate to Admin Center ➤ Monitor Job and verify that the rules were imported. Figure 12-1 shows a sample screenshot from a prior import. Instances created more recently may not show the import processes, since they were pre-delivered.

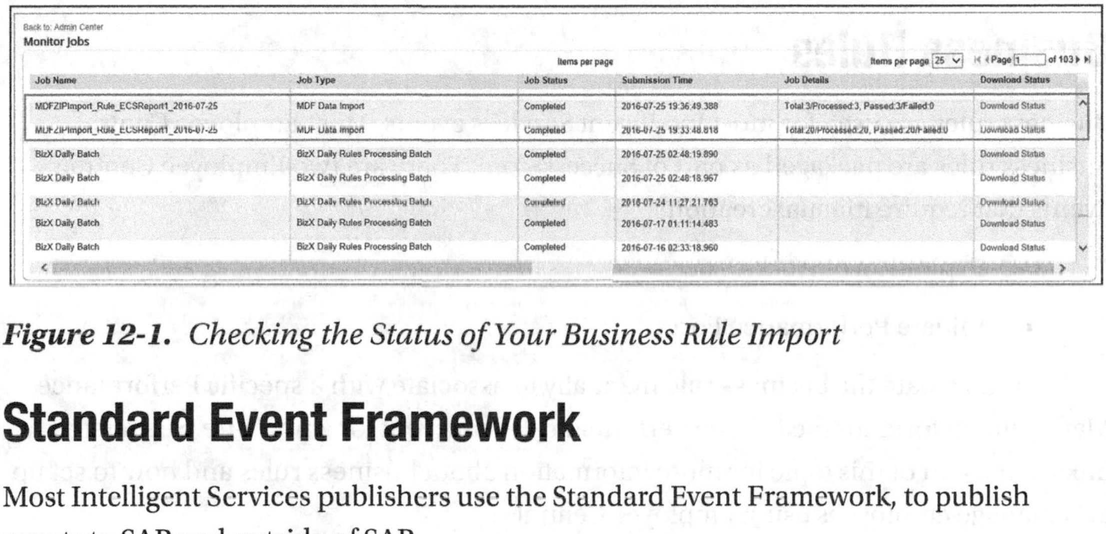

Job Name	Job Type	Job Status	Submission Time	Job Details	Download Status
MDFZIPImport_Rule_ECSReport1_2016-07-25	MDF Data Import	Completed	2016-07-25 19:36:49.388	Total:3/Processed:3, Passed:3/Failed:0	Download Status
MDFZIPImport_Rule_ECSReport1_2016-07-25	MDF Data Import	Completed	2016-07-25 19:33:48.818	Total:20/Processed:20, Passed:20/Failed:0	Download Status
BizX Daily Batch	BizX Daily Rules Processing Batch	Completed	2016-07-25 02:48:19.890		Download Status
BizX Daily Batch	BizX Daily Rules Processing Batch	Completed	2016-07-25 02:48:18.967		Download Status
BizX Daily Batch	BizX Daily Rules Processing Batch	Completed	2016-07-24 11:27:21.763		Download Status
BizX Daily Batch	BizX Daily Rules Processing Batch	Completed	2016-07-17 01:11:14.483		Download Status
BizX Daily Batch	BizX Daily Rules Processing Batch	Completed	2016-07-16 02:33:18.960		Download Status

Figure 12-1. *Checking the Status of Your Business Rule Import*

Standard Event Framework

Most Intelligent Services publishers use the Standard Event Framework, to publish events to SAP and outside of SAP.

Intelligent Services uses existing ODATA APIs to fetch additional information about the event and comprises records of the same type.

Some events only publish to other internal SAP SuccessFactors subscribers, or to external third-party applications, or both.

Service Event Bus

The Service Event Bus Architecture allows the event message to flow from publisher to subscribers. The Service Event Bus utilizes ODATA database tables that contain connected entities that allow internal SAP SuccessFactors applications to receive notifications from publishers. This architecture framework allows external third-party applications to subscribe to Intelligent Services.

Most Intelligent Services events publish to subscribers using the Service Event Bus. There are events available as a result of integration work between Employee Central and SAP Jam or from form routing workflow using Performance Management forms for a new hire.

Business Rules

Business rules are used for most Intelligent Services events. Most Employee Central business rules are packaged as part of SuccessStore. There are two Employee Central events that require manual creation:

- Expiration of Work Order

- Initiate Performance Form

You can create the business rule manually to associate with a specific Performance Management form for the Initiate Performance Form event for a new hire. See related links at the end of this topic for more information about business rules and how to set up and manage employees using Employee Central.

Note   Business Rules cannot be used currently for recruiting.

Event Structure and Payload for Recruiting

The official documentation shows the event structure and payload for the events as shown in Table 12-2. The source for all SAP SuccessFactors events for recruiting is MODULE_RCM.

Table 12-2. Recruiting Events and Payloads

Event Name	Event Description	Event Type	ODATA Entity Type	Entity Keys and Data Type	Parameter Keys and Data Type
Update of Job Requisition	This event is raised when an approved or closed job requisition is updated.	JobRequisition.Updated	JobRequisition	jobReqId (string)	updatedFieldsList (string) isOnboardingInitiated (string)
Update of Job Application	This event is raised when a job application is updated and the applicant is an applied state.	JobApplication.Updated	JobApplication	applicationId (string)	updatedFieldsList (string) isOnboardingInitiated (string)
Approval of Offer Detail	This event is raised after the final step in the job offer detail.	JobOffer.Updated	JobOffer	offerApprovalId (string)	applicationId (string) isOnboardingInitiated (string)
Update of Job Application Status	This event is raised when an application status is changed.	JobApplication. StatusUpdated	JobApplication	applicationId (string)	candidateId (string)

(continued)

Table 12-2. (*continued*)

Event Name	Event Description	Event Type	ODATA Entity Type	Entity Keys and Data Type	Parameter Keys and Data Type
Update of Candidate Profile	This event is raised when an update is made to the candidate profile.	CandidateProfile.Updated	Candidate	candidateId (string)	None
Initiate Job Posting	This event is raised when a job is posted to the external career site.	JobPosting.Change	JobRequisitionPosting	jobPostingid (string)	jobReqId (string) jobPostingType (string)
Update Candidate's Country to Russia	This event is raised when a candidate's country is changed to Russia.	CandidateProfileUpdated. CountryUpdated	Candidate	candidateId (string)	isCountryChangeEvent (string)

Chapter 1 covered basic aspects of ODATA APIs. You can also get the events in a customer instance (or any instance for that matter) by simply executing GET <<endpoint>>odata/v2/getExtEventMetaDataDefinition; however, keep in mind that there's no filter. You will get all events in an instance.

Subscribing to an Event

The first step in setting up an integration with Intelligent Services is identifying which event to subscribe to. To do this, navigate to Admin Center ➤ ISC (Intelligent Services Center). Figure 12-2 shows the seven recruiting events described in this document. The current section will use "Update of Job Requisition" as an illustration throughout.

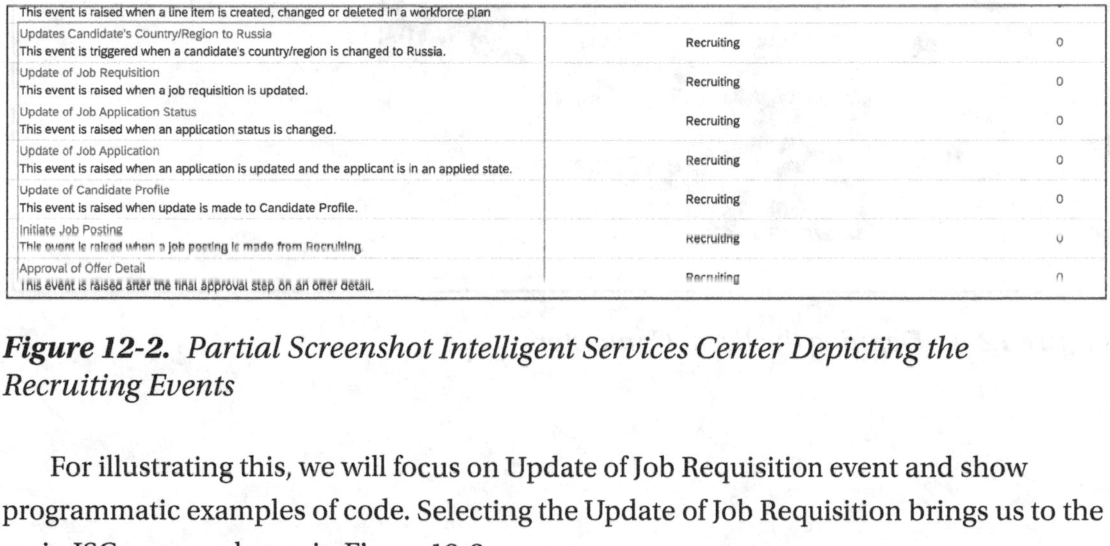

This event is raised when a line item is created, changed or deleted in a workforce plan		
Updates Candidate's Country/Region to Russia. This event is triggered when a candidate's country/region is changed to Russia.	Recruiting	0
Update of Job Requisition This event is raised when a job requisition is updated.	Recruiting	0
Update of Job Application Status This event is raised when an application status is changed.	Recruiting	0
Update of Job Application This event is raised when an application is updated and the applicant is in an applied state.	Recruiting	0
Update of Candidate Profile This event is raised when update is made to Candidate Profile.	Recruiting	0
Initiate Job Posting This event is raised when a job posting is made from Recruiting.	Recruiting	0
Approval of Offer Detail This event is raised after the final approval step on an offer detail.	Recruiting	0

Figure 12-2. *Partial Screenshot Intelligent Services Center Depicting the Recruiting Events*

For illustrating this, we will focus on Update of Job Requisition event and show programmatic examples of code. Selecting the Update of Job Requisition brings us to the main ISC page as shown in Figure 12-3.

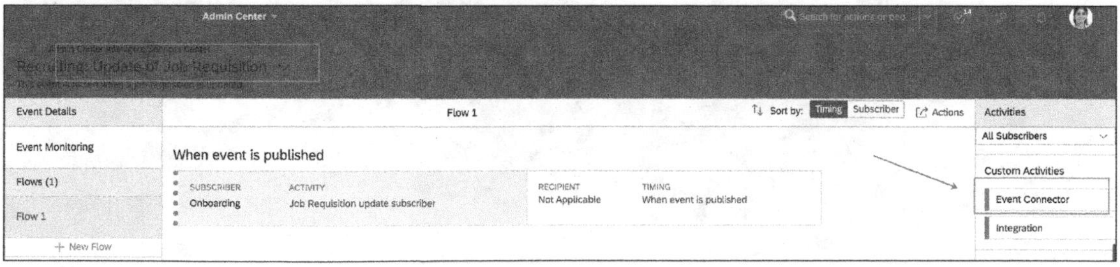

Figure 12-3. *Details of the Update of Job Requisition Event*

Click Event Connector to bring up the page where we will define the connector. Note that Chapter 10 shows another example using Update of Job Application Status. Click the Event Connector in Figure 12-3 to display the Event Connector. Figure 12-4 shows the Event Connector.

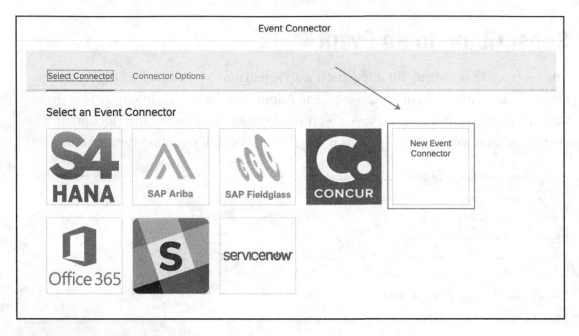

Figure 12-4. *Defining the Event Connector*

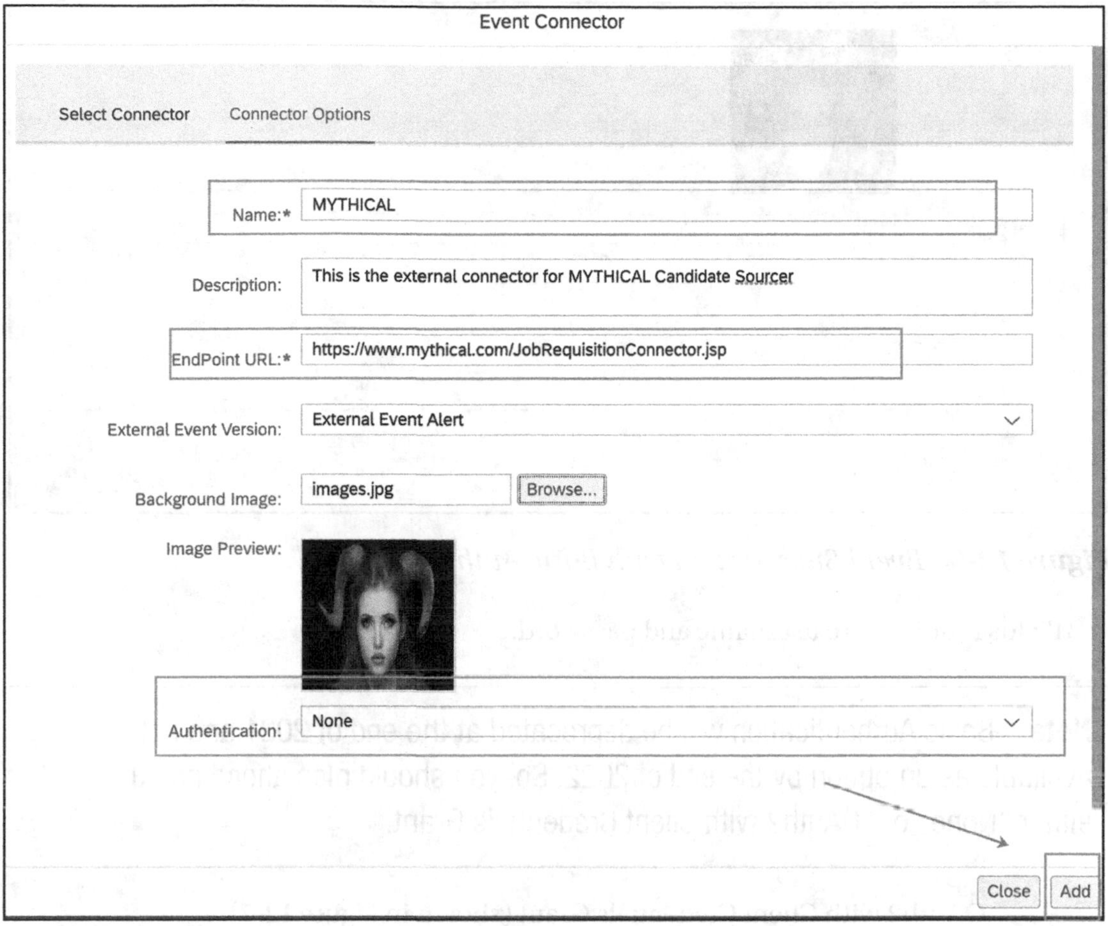

Figure 12-5. *Defining the MYTHICAL Event Connector*

Figure 12-5 shows a completed event connector without any authentication. Please note that the Name and Endpoint URL are required fields and the Endpoint URL is fictitious.

SAP SuccessFactors offers three authentication options:

– None (as shown in Figure 12-5)

– Basic (shown in Figure 12-6)

Figure 12-6. *Event Subscription with Basic Authentication*

Fields required are username and password.

Note   Basic Authentication will be deprecated at the end of 2021 and not available as an option by the end of 2022. So, you should plan ahead and use either "None" or "OAuth2 with Client Credentials Grant."

–   OAuth2 with Client Credentials Grant (shown in Figure 12-7).

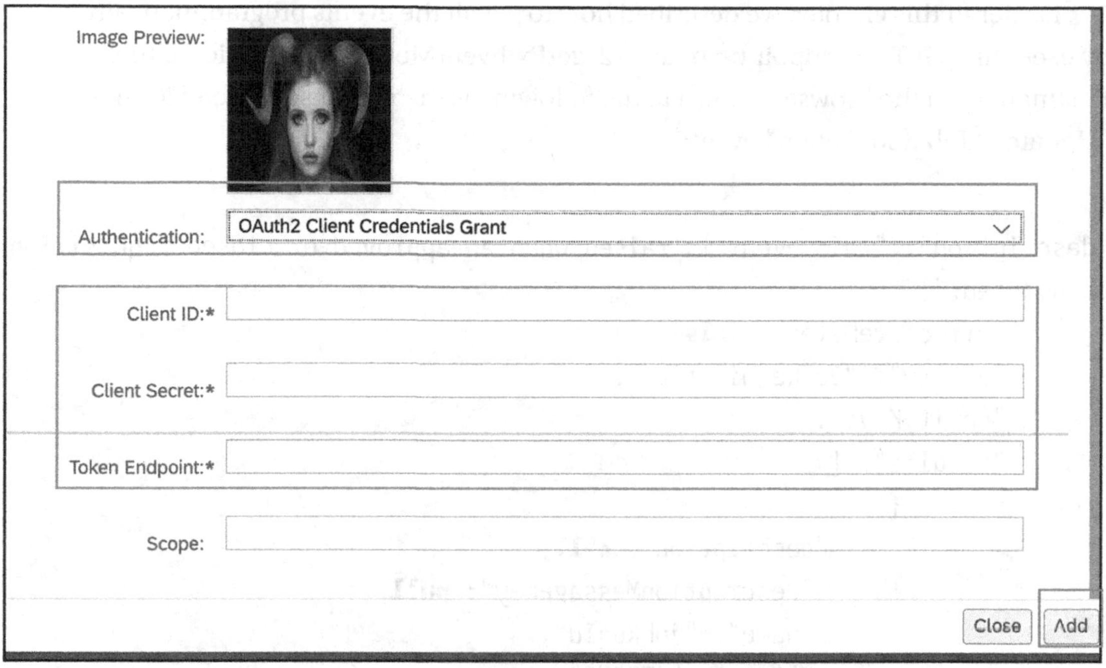

Figure 12-7. *Event Subscription with OAuth2 with Client Credentials Grant*

The required fields are Client ID, Client Secret, and Token Endpoint, while Scope is optional. These properties are defined on the third-party platform. Regardless of the authentication option selected, you have to click "Add" to register the subscription. The final step is to ensure that your flow is saved as in Figure 12-8.

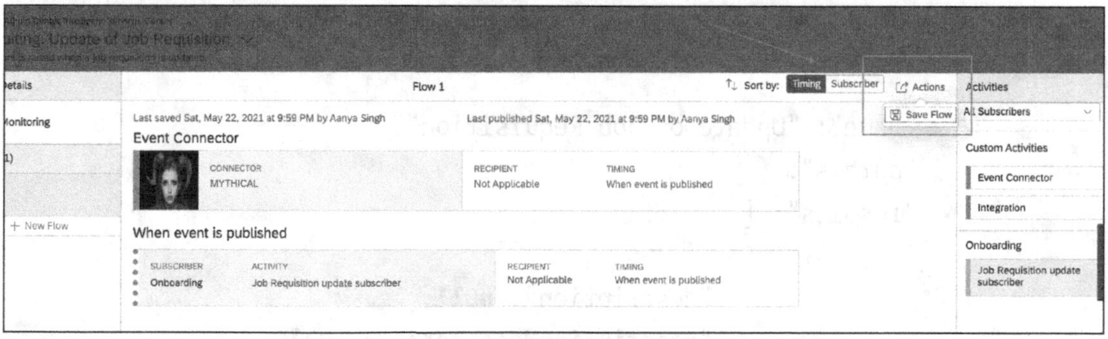

Figure 12-8. *Saving Your Final Flow*

Earlier in this chapter, we described how to get all the events programmatically by executing GET <<endpoint>>odata/v2/getExtEventMetaDataDefinition either in Postman or on the browser. Shown in the following is the response payload for the "Update of Job Requisition" event:

```
{
"description": "This event is raised when an approved or closed requisition
is updated.",
      "effectiveDated": false,
      "entity": "JobRequisition",
      "entityKeys": {
      "results": [
            {
                  "description": null,
                  "descriptionMessageKey": null,
                   "name": "jobReqId",
                    "type": "STRING"
            },
            {
                     "description": null,
                  "descriptionMessageKey": null,
                  "name": "startDate",
                  "type": "DATE"
            }
      ]
      },
         "name": "Update of Job Requisition",
         "params": {
         "results": [
               {
                     "description": null,
                     "descriptionMessageKey": null,
                     "hasValueAlways": false,
                     "name": "updatedFieldsList",
                     "type": "STRING"
               },
```

```
                {
                    "description": null,
                    "descriptionMessageKey": null,
                    "hasValueAlways": false,
                    "name": "isOnboardingInitiated",
                    "type": "STRING"
                }
            ]
        },
            "publisher": "Recruiting",
          "topic": "com.successfactors.recruiting.JobRequisition.
          Updated",
          "type": "JobRequisition.Updated"
}
```

Once your subscription is configured and the connection to the external system is successful, the payload is delivered via a SOAP message shown as follows:

```
<?xml version="1.0" encoding="UTF-8" standalone="yes"?>
<ns6:ExternalEvent xmlns:ns6="http://notification.event.successfactors.
com" xmlns:ns5="http://schemas.xmlsoap.org/soap/envelope/" xmlns:ns7="com.
successfactors.event.notification" xmlns:ns2="com.successfactors.alert"
xmlns:ns4="http://alert.successfactors.com" xmlns:ns3="http://www.boomi.
com/connector/wss">
    <ns6:externalEventMeta>
        <ns6:externalEventId>b772edb5-9f59-4b53-8595-6e45d08c0756</
        ns6:externalEventId>
        <ns6:type>com.successfactors.recruiting.JobRequisition.Updated</
        ns6:type>
        <ns6:publishedAt>1621793760938</ns6:publishedAt>
        <ns6:publishedBy>sfadmin</ns6:publishedBy>
        <ns6:repost>false</ns6:repost>
    </ns6:externalEventMeta>
    <ns6:events>
        <ns6:event>
            <ns6:eventId>fa477244-08c2-4293-aa65-2bad255b75d9</ns6:eventId>
            <ns6:entityType>JobRequisition</ns6:entityType>
```

```
        <ns6:publishedAt>1621793760929</ns6:publishedAt>
        <ns6:publishedBy>sfadmin</ns6:publishedBy>
        <ns6:repost>false</ns6:repost>
        <ns6:entityKeys>
            <ns6:entityKey>
                <name>jobReqId</name>
                <value>2561</value>
            </ns6:entityKey>
        </ns6:entityKeys>
        <ns6:params>
            <ns6:param>
                <name>updatedFieldsList</name>
                <value>workHours[en_US], experienceReq</value>
            </ns6:param>
            <ns6:param>
                <name>isOnboardingInitiated</name>
                <value>false</value>
            </ns6:param>
        </ns6:params>
      </ns6:event>
    </ns6:events>
</ns6:ExternalEvent>
```

Notes on the SOAP payload:

1. The payload is obtained from the external system after the event was successfully retrieved. In this example, the external system was SAP's Business Technology Platform (BTP), a middleware.

2. The only namespace applicable for external entities is ns7. However, it's best practice to ensure that the third-party software is set to ignore namespaces.

3. externalEventId is an attribute that applies to the whole data center while eventId is specific to the current instance.

4. publishedAt is expressed in Unix epoch time relative to
    UTC. Notice that there are two publishedAt properties separated
    by a few milliseconds. Using `https://currentmillis.com` as a
    reference, we can determine that 1621793760929 translates to
    "Sun May 23 2021 18:16:00" in UTC date and time and "Sun
    May 23 2021 11:16:00" in local time, Pacific Daylight Saving Time.

5. publishedBy is the userId of the user in the current customer
    instance. In this example, it's twalker (Tessa Walker) in a sales
    demo instance.

6. The updateFieldsList parameters collection indicates that the
    properties "workHours[en_US]" and "experienceReq" have
    changed on the "entityType" of JobRequisition. You can use these
    parameters to determine the next course of action.

7. The "type" property in the SOAP payload indicates the underlying
    event in question, JobRequisition.Updated.

Troubleshooting Your Integration

Once your integration is set up, the next logical step is to figure out if your integration is
working. In order to do that, access Admin Center ➤ ISC (Intelligent Services Center).
You will see the event raised within SuccessFactors.

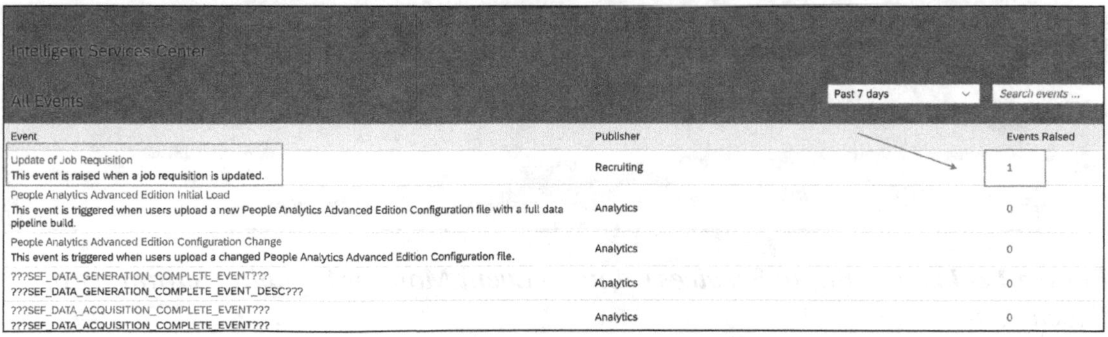

Figure 12-9. *Intelligent Services Center to Show the Raised Event for Update of Job
Requisition*

Click the link that shows the number under the "Events Raised" column. Figure 12-10 shows such an example.

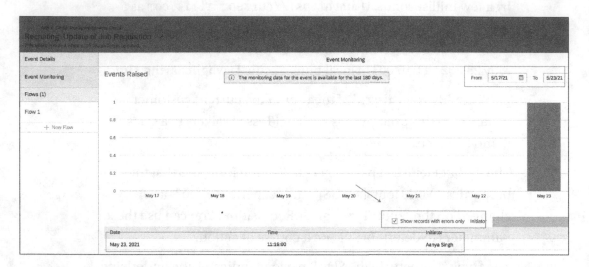

Figure 12-10. *Intelligent Services Center Screenshot Showing the Event Raised*

You may optionally choose to view only those events that were unsuccessful by clicking the "Show records with errors only" checkbox as shown in Figure 12-10. Please note that you can use the date filter to narrow or expand your results. The executions are limited to the past 180 days. Now click the event you're interested in to view details as shown in Figure 12-11.

Figure 12-11. *Intelligent Services Center – Event Monitor Showing Summary of Execution*

Note that the event for the MYTHICAL connector shows a Failed Status with a pop-up to show details of the execution. You may choose to click directly the pop-up for "Show Details" or access the Event Notification Audit Log as shown in Figure 12-15. Figure 12-11 is useful if you have very few events and you want to quickly view the details.

The following payload shows the actual request sent by SuccessFactors to the endpoint defined in the setup, namely, `www.mythical.com/JobRequisitionConnector.jsp`. Figure 12-12 shows an example of the request SOAP payload sent by SuccessFactors to the external entity MYTHICAL defined earlier.

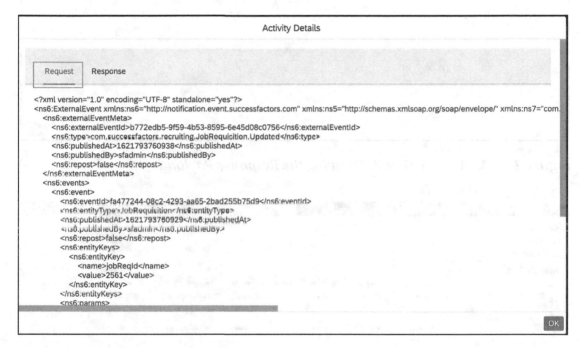

Figure 12-12. *Activity Details Showing the Request Payload*

Figure 12-13 shows the response generated by the SuccessFactors while it attempts to connect to the endpoint defined by the MYTHICAL connector, `www.mythical.com/JobRequisitionConnector.jsp`.

Activity Details

Request   Response

```
---------------------------------------------------------------
ExecuteTime=[1]
EndPointURL=[https://www.mythical.com/JobRequisitionConnector.jsp]
Exception=[Could not send Message.]
StackTrace=[javax.xml.ws.WebServiceException: Could not send Message.  at org.apache.cxf.jaxws.DispatchImpl.mapException(DispatchImpl.java:267)
...more
Caused by:
org.apache.cxf.transport.http.HTTPException: HTTP response '404: Not Found' when communicating with https://www.mythical.com/JobRequisitionConn
...more
].
```

OK

Figure 12-13. *Activity Details Showing the Response Payload*

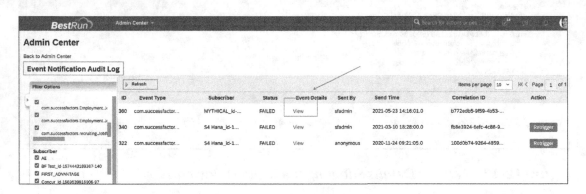

Figure 12-14. *Landing Page for Event Notification Audit Log*

If you have a lot of events, it may be useful to access Admin Center ➤ Event Notification Audit Log as in Figure 12-14.

The events are listed in the order of execution. If there are a lot of executions, you can use the Filter Options shown in Figure 12-14 to narrow down the events. In the example, the relevant event is com.successfactors.recruiting.JobRequisition.Updated. You may also click Refresh to bring up the latest executions. You can click the "View" hyperlink under the "Event Details" column to see the details of the execution. The request and response payloads are the same if launched via Figure 12-11 by clicking "Show Details" or by clicking "View" column as in Figure 12-14. The request and response payloads are illustrated in Figures 12-12 and 12-13.

If you're certain that you've raised an event, but for some reason, you don't see in the Event Monitor as in Figure 12-10 or 12-14, the option is to raise a support ticket on behalf of the customer.

Best Practices to Follow for Intelligent Services

Similar to other software solutions, use of Intelligent Services always presents challenges and compromises. It requires some past experience and some analysis to determine how best to leverage Intelligent Services. The following are a few things to keep in mind while leveraging Intelligent Services. By no means is this an exhaustive list, and we welcome your input from your experience.

- The number of events raised is entirely unpredictable. If there are an unusual number of events, the Service Event Bus could get clogged, and the only resort is to open a support ticket on behalf of the customer. This is a rare occurrence and shouldn't normally affect most integrations. There are customer instances with 100,000 transactions a day and 8 million candidates and 15 million job applications and nearly 300,000 job requisitions. As a general rule, "Update of Job Application" will be in multiples of events for "Update of Job Requisition."

- Once an event is subscribed to, there's no control over the number of events raised or the amount of over the network traffic occurring between SuccessFactors and the third party.

- Depending on the customer and usage patterns, you could have thousands or even hundreds of thousands of events raised during a day. You have to consult with the customer on their usage pattern whether it is really necessary to have near real-time integration.

- Each event will raise a distinct transaction over the network, and you should be aware of such situations.

- Use caution when deciding between "Update of Job Application" and "Update of Job Application Status" events since both events are based on the JobApplication entity. If you're interested only in the status change, you should strongly consider using "Update of Job Application Status" event. An example of this is available in Chapter 12 about Background Checks.

- Use scheduled processes as described in Chapter 7 regarding Integration Center for large data extracts.

- As mentioned earlier in this chapter, SuccessFactors will be deprecating Basic Authentication by end of 2021 and entirely unavailable by end of 2022. So, every effort should be made to migrate existing event integrations employing Basic Authentication. In the worst-case scenario, you can use None as the authentication, but you have to evaluate the customer's needs.

- Use "Update of Candidate Profile" event sparingly. If you use this event, you will have more transactions than are needed, and your software should be able to discard those that are not needed.

Combining Intelligent Services with Integration Center

As mentioned earlier in this chapter, Intelligent Services will produce an asynchronous SOAP message for each event raised. If you need more details about the underlying entity that raises an event, you need to make explicit ODATA API calls. Another way to take advantage of the in-built tools is to use Integration Center that allows you to enhance the payload and deliver the payload in JSON or pure XML or even SFTP if that's the only way to consume information from SuccessFactors. Taking the earlier example of MYTHICAL, now let's define an integration template in the Integration Center. See Figure 12-15 on where to add this integration.

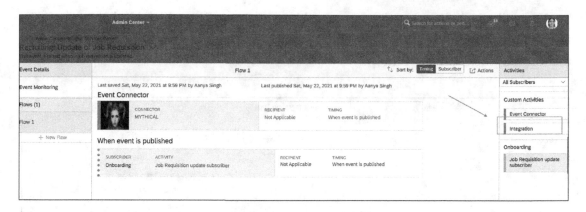

Figure 12-15. *Associating an Integration with an Event*

When you click Integration as in Figure 12-15, you will see a pop-up indicating that you don't have an integration defined. Once you acknowledge the pop-up, you will see the pop-up as in Figure 12-16 and you can now create a new integration.

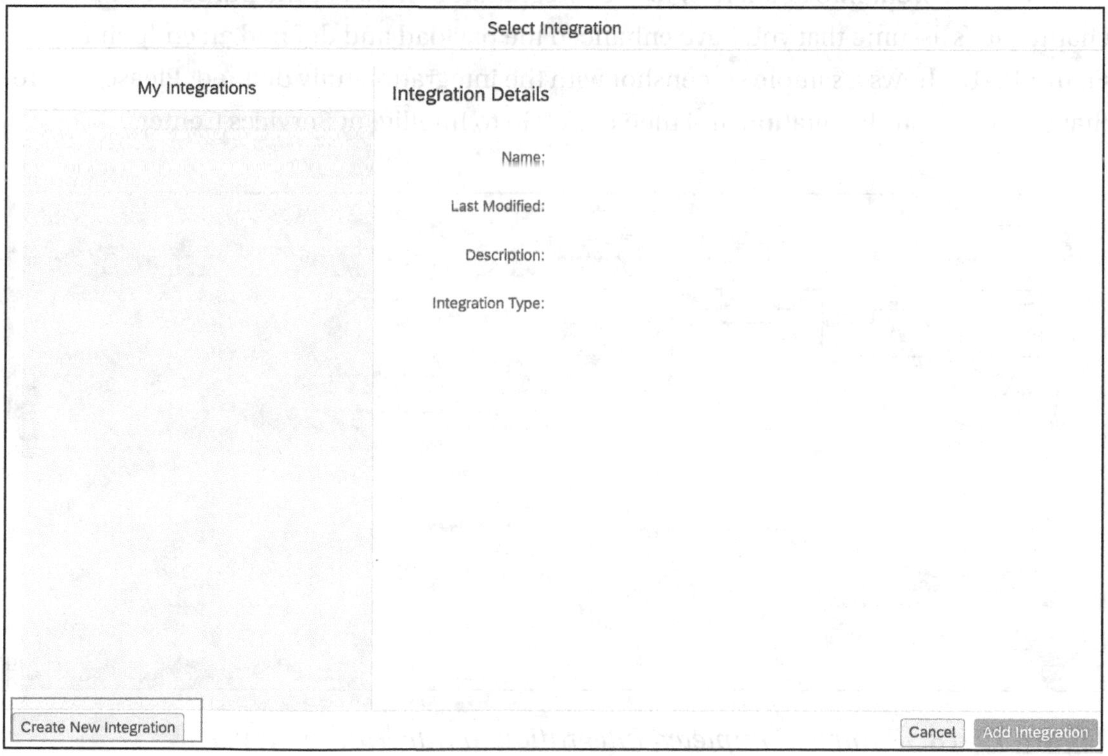

	Select Integration
My Integrations	Integration Details
	Name:
	Last Modified:
	Description:
	Integration Type:
Create New Integration	Cancel Add Integration

Figure 12-16. *Creating a New Integration Pop-Up*

You will now be presented a pop-up asking you to confirm which kind of an integration you want to build. Figure 12-17 shows an example with REST and JSON as the format. Click "Create" button.

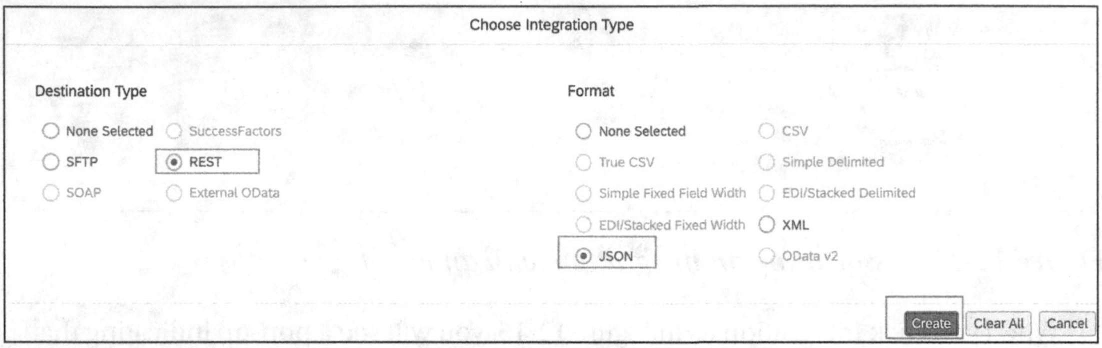

Choose Integration Type

Destination Type

○ None Selected ○ SuccessFactors
○ SFTP   ⦿ REST
○ SOAP   ○ External OData

Format

○ None Selected   ○ CSV
○ True CSV   ○ Simple Delimited
○ Simple Fixed Field Width ○ EDI/Stacked Delimited
○ EDI/Stacked Fixed Width   ○ XML
⦿ JSON   ○ OData v2

[Create] [Clear All] [Cancel]

Figure 12-17. *Integration Center Pop-Up to Define Format*

Chapter 7 went into detail how to use Integration Center. For the purposes of this chapter, let's assume that you have enhanced the payload and defined an endpoint. Figure 12-18 shows a sample screenshot with the integration fully defined. Please ensure that you save your integration and then click "Go to Intelligent Services Center."

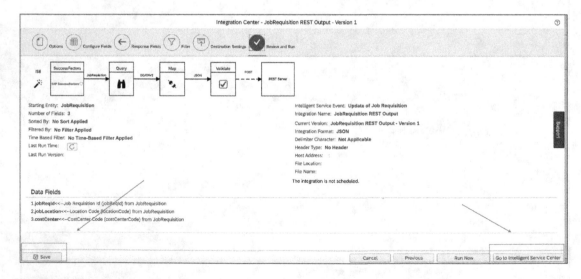

Figure 12-18. *Sample Completed Integration in Integration Center*

Now click "Integration" as shown in Figure 12-15. This time around, the screen will look the same as Figure 12-16, but with the newly saved integration populated. See Figure 12-19 for an example.

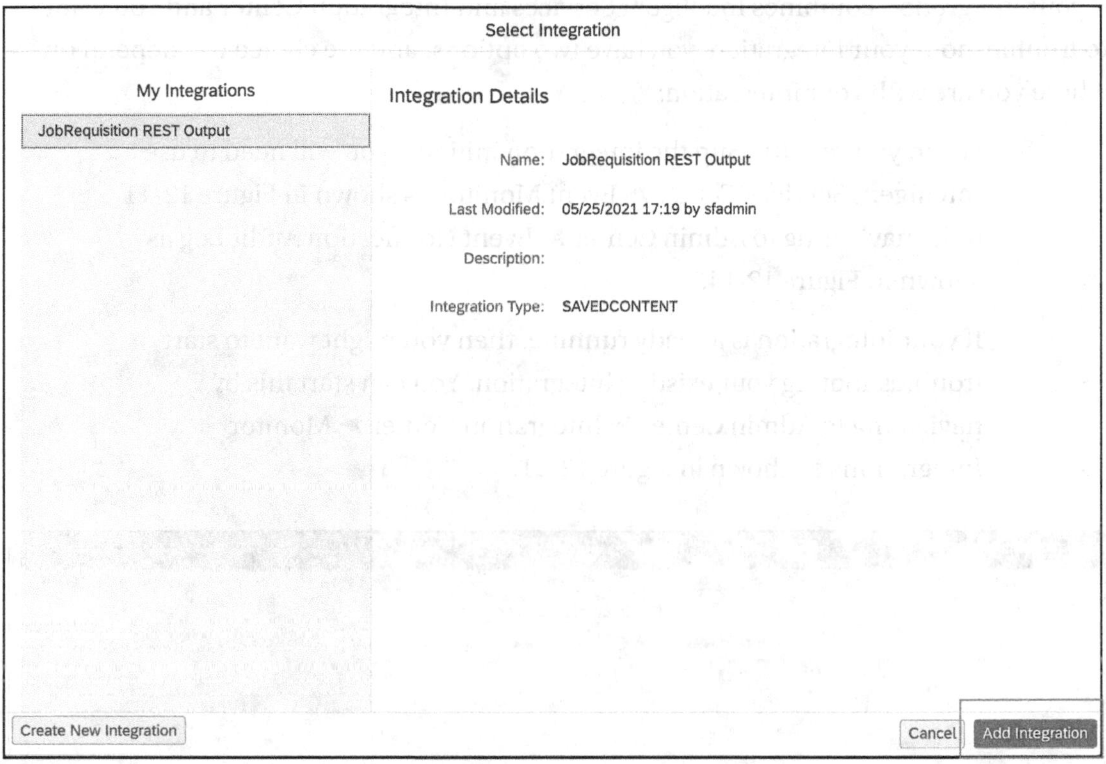

Figure 12-19. *Pop-Up Showing Previously Added Integration*

Once you've added the integration, ensure that you click "Save Flow" as shown in Figure 12-20.

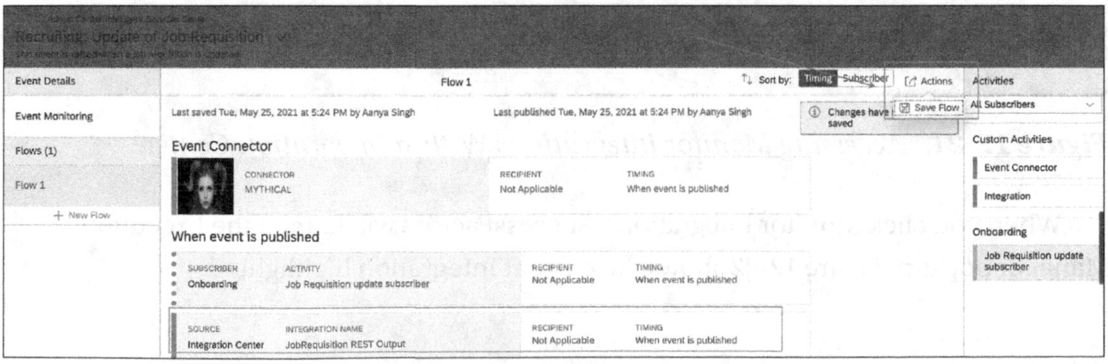

Figure 12-20. *Saving Your Integration Flow*

Troubleshooting Your Combined Integration Using Intelligent Services and Integration Center

If your integration combines Intelligent Services and Integration Center and you want to troubleshoot your integration, you have two options, and the choice will depend on where you are with your integration:

- When you're setting up the integration initially, you will need to use Intelligent Services Center ➤ Event Monitor as shown in Figure 12-11 or by navigating to Admin Center ➤ Event Notification Audit Log as shown in Figure 12-14.

- If your integration is already running, then you might want to start troubleshooting your existing integration. You can start this by navigating to Admin Center ➤ Integration Center ➤ Monitor Integrations as shown in Figure 12-21.

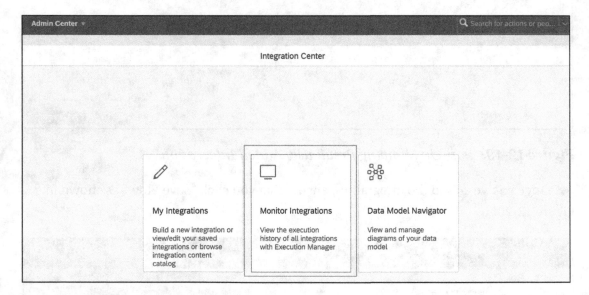

Figure 12-21. *Accessing Monitor Integrations Within Integration Center*

When you click Monitor Integrations, SuccessFactors will launch the Executions Manager pop-up. Figure 12-22 shows the current integration highlighted.

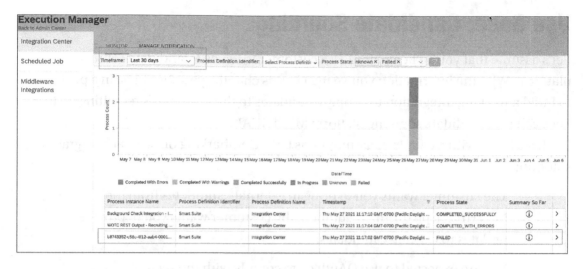

Figure 12-22. *Screenshot Showing the Execution Manager Within Integration Center*

You will notice that the integration shows status as failed. This is because the fictitious endpoint we chose in Figure 12-18 doesn't really exist and the connection fails. Figure 12-22 also shows a successful completed integration using Background Checks in Chapter 10. Click the arrow to see additional details of the execution as shown in Figure 12-23. You can click the download icon to export the results in a CSV file. If you wish to share the details of your results in a support ticket, it's useful to attach the error log.

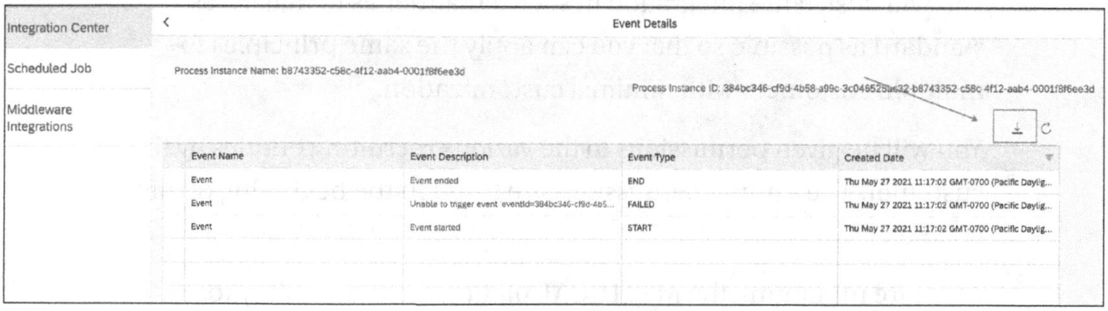

Figure 12-23. *Sample Detail About the Event Trigger*

Now that you have seen how to set up the integration using Intelligent Services and Integration, let's examine a practical use case to walk through the process of setting up such an integration in a customer instance.

Use Case: Candidate Sourcing

Let's assume that you've been hired by a customer to act as a candidate sourcing platform. With the information you've read in this chapter, you should be in a position to build a working integration to supply candidates to the customer's recruiting system. Let's call this candidate sourcing platform MYTHICAL.

Let's begin with a couple of assumptions before embarking on such an integration solution:

- The customer wants you to bring in external candidates from the MYTHICAL platform. In addition, the customer wants you to provide updates to existing candidates.

- You are expected to use OAuth2.0 to comply with the security requirements of the customer and SAP SuccessFactors.

- The job requisitions should be posted externally first before you can source candidates.

- Leveraging ODATA APIs, you will create candidate records in SuccessFactors in addition to any supporting attachments such as resumes and cover letters.

- The candidates should be aware of any new job requisitions posted by the customer in near real time.

- You will design the integration in such a manner as to make it as standard as possible so that you can apply the same principles to multiple customers with minimal customization.

- You will be given permissions to the various recruiting entities by the customer. The sample permissions are listed in the beginning of this chapter.

- At a bare minimum, the MYTHICAL platform has the ability to receive SOAP messages from SuccessFactors. A nice feature to have will be the ability of the platform to receive JSON messages from SuccessFactors (in case you are going to combine Intelligent Services with Integration Center).

- The MYTHICAL platform also has the ability to invoke ODATA APIs against the customer's Recruiting APIs.

- If for some reason, there's a limitation on the MYTHICAL platform to consume events, you should make accommodations to address this limitation with the customer.

- For the purposes of this chapter, let's assume that the candidate sourcing is limited to North America.

The steps involved in building a solution can follow the industry-accepted approaches to software development life cycle. Typically, the steps needed are Requirements Gathering and Analysis, Design, Implementation, Testing, Deployment, and Maintenance. A detailed treatment of the steps is outside the scope of this chapter. Given the assumptions and the customer's requirements, you should be in a position to complete the first step, namely, Requirements Gathering and Analysis.

In order to complete the Requirements Gathering and Analysis, you may need to have several discussions with the prospective customer before you embark on the next step – Design. What objects/entities will you need access to? At a bare minimum, you need access to the following APIs, Candidate, JobRequisition, and JobApplication. Chapter 2 covered the basics of working with ODATA. Please refer to that chapter for more information.

Candidate Sourcing – Process Integration Flow:

1. Customer creates job requisitions, updates them, and posts them externally for sourcing.

2. MYTHICAL platform will initially get all relevant job requisitions en masse. The amount of information about the job requisition will depend on the requirements of the MYTHICAL platform and typically a fraction of the information maintained with the SuccessFactors job requisition.

3. Customers will periodically update the job requisitions.

4. MYTHICAL platform will be notified of the updates to job requisitions by subscribing to the Update of Job Requisition Intelligent Services event and will subsequently update the job requisitions on the MYTHICAL platform.

5. MYTHICAL platform will manage candidate acquisition and will submit candidates to SuccessFactors at an appropriate time using the Candidate OData API, optionally with the resume and cover letters. SuccessFactors Recruiting will return the candidate ID as part of the response to the candidate creation operation. MYTHICAL platform will notify the recruiter that candidates have been created either periodically or on an ad hoc basis. This should be discussed with the customer as early in the integration design.

6. The recruiter will invite the candidate created by MYTHICAL to apply to a job requisition.

7. Once the candidate successfully applies to a job requisition, a record for the JobApplication ODATA entity will be created.

8. MYTHICAL platform can optionally forward the candidate to a particular job requisition using the JobReqFwdCandidates API or defer the decision to the customer recruiter.

9. MYTHICAL can query the Candidate API to find out which job applications have been created by using the Candidate. jobsApplied navigation property.

10. The customer will update the job application status periodically and will raise the Update to Job Application Status event, and the MYTHICAL platform will be notified of these updates.

11. MYTHICAL platform can optionally notify the candidates of all status changes to a job application or choose only certain ones as required.

Design Considerations:

Of course, designing such an integration will largely depend on the capabilities of the MYTHICAL platform. Some considerations include

– Does the integration have to be near real time? The answer to this question largely depends on the capability of the MYTHICAL platform to handle Intelligent Services events. If there is no option to react immediately to an Intelligent Services event, the only option is then to invoke ODATA API calls to SuccessFactors from the MYTHICAL platform.

- Can the MYTHICAL platform receive only asynchronous SOAP messages? If that's the case, then the Integration Center is out of consideration. This implies that the troubleshooting will be limited to Admin Center ➤ Event Notification Audit Log.

- If the MYTHICAL platform can receive and react to synchronous JSON messages, then the event can be combined with Integration Center and deliver a synchronous JSON message. This implies that the troubleshooting can be done at two locations in SuccessFactors, Integration Center ➤ Monitor Message and Admin Center ➤ Event Notification Audit Log.

- If the MYTHICAL platform is interested only in certain job application status messages, you can include a filter on the Integration Center template for this integration. In addition, you can enhance the payload to only choose those fields/properties that are relevant to the platform.

- Is this integration meant to address the needs of one customer or many? If the need is for one customer, you can just use Intelligent Services ➤ Event Connector. However, that means you cannot take advantage of Integration Center's feature to enhance the payload. If you're designing the integration for multiple customers, it's advantageous to leverage the Integration Center and include a static property to identify the customer's instance.

- If you're designing this integration for multiple customers, consider the option of saving the Integration Center Definition (ICD) template and sharing the ICD template with customers ahead of time. This will save implementation time for individual customers.

- If you're designing this integration for customers in other geographic regions, be aware that SuccessFactors has several data centers, each with its own base URLs and API servers.

- You will want to take into account that each customer will have more than one instance on different data centers, Production, Preview, and Development. Do not assume that the integration will be designed with only one URL/API server in mind.

– Based on the Process Integration previously described, MYTHICAL will be interested in two Intelligent Services events, Update of Job Requisition and Update of Job Application Status.

– Since not all event notifications and/or updates will be relevant to MYTHICAL, you should accommodate the reason for ignoring an update.

– While invoking APIs to retrieve job requisitions or candidates, please be sure not to do a full load of all job requisitions and candidates each time. It's preferable to use the filter property lastModified on the JobRequisition and lastModifiedDateTime on the Candidate entity to obtain the delta load.

– For creating candidates via ODATA API, consider using CandidateLight instead of Candidate entity, because it has fewer required fields and is almost always the same across customers. Remember that the recruiting configurations vary vastly by customers and can complicate your integrations especially if you're trying to design this integration for multiple customers.

Conclusions

Upon completion of this chapter, you should be well versed with the Intelligent Service Events published by SuccessFactors Recruiting. Following the design considerations, you should be able to successfully architect a solution for external candidate sourcing for a customer or multiple customers.

CHAPTER 13

Hiring Integration with Non-SAP Systems

We've made it to the end of the recruiting process. Now there's one final step left – hire the candidate! However, if we are not using the standard integrations with SAP SuccessFactors Employee Central or SAP HCM On-Premise, how can we integrate the hiring process? In this chapter, we explore the relevant data and tools available to set up this automation.

Example Integrations and Automations at Hire

The prime example of a hiring integration is to pass data about a candidate to an HRIS system. In the most common scenario, once you decide to hire a candidate, an HRIS system requires basic information like name, tax ID, and address that has already been collected from the candidate. It therefore makes sense to send this information automatically rather than having the new hire and/or a human resources administrator rekey the information. In addition, information like salary and bonus may need to be sent from recruiting to payroll.

Note *Consult with your local country legal council and HR administrators on the most appropriate stage to collect tax ID – in some countries, the tax ID is collected during the recruiting process where in others it is not.*

Often, an onboarding system lies between the recruiting and core HRIS/payroll system(s). These onboarding systems can provide an intermediary between the HRIS and payroll systems where prospective new hires complete paperwork such as government forms and reviewing and signing required company policies. This helps to

331

© Anand 'Andy' Athanur, Mark Ingram and Michael A. Wellens 2022
A. A. Athanur et al., *Innovative SAP SuccessFactors Recruiting*, https://doi.org/10.1007/978-1-4842-7425-5_13

ensure employees do not spend their first day on payroll filling out paperwork and gives systems and HR administrators a chance to work with the potential hire information prior to committing them as hired in the HRIS/payroll system(s) (which also helps to soften the blow if a potential hire does not show on day one).

Outside of onboarding, HRIS, and payroll systems, you may also wish to trigger other events that help with onboarding and hiring activities. For example, we can send information to a ticketing system and automatically open a ticket requesting equipment such as a laptop or uniform. You may also have third-party vendors that require basic information about incoming employees for training events. We can also send information to set up system user IDs for learning systems or role-specific specialized systems that are needed to make employees productive on day one.

Clearly, there are a number of very practical time and cost-saving business scenarios! It is therefore important to understand what data is available for us to use to help us take advantage of these scenarios. Let's take a look in the next section!

Relevant Data

In this section, we quickly review the most commonly accessed data for hiring integrations and automations and provide guidance on when each object is typically used. Figure 13-1 illustrates a high-level structure of the most commonly used data in the aforementioned business scenarios. It is important to understand what information is stored where and how the objects relate to one another so that any interfaces or automations you create grab data from the most relevant object. For example, there is often redundant data shared between these objects, so which one should you pull data from in what scenario? We will go into more detail on each of these objects in the following sections.

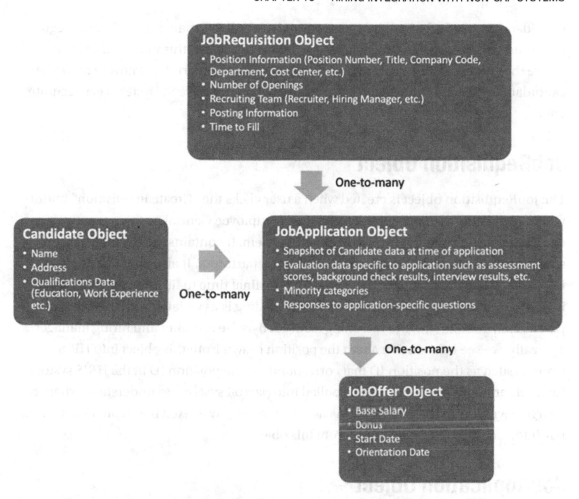

Figure 13-1. *High-Level Structure of Data Objects Most Commonly Relevant for Hiring Integrations and Automations*

Candidate Object

The Candidate object is created any time a candidate creates an account on the external career site, an internal SuccessFactors user visits the internal career site, or a recruiter manually creates a candidate (or an interface you built creates one!). It contains basic information about the person such as first and last name, address, and also qualification data such as education, prior work experience, or any other background elements that have been configured. Most often, you will want to pull any personal information about a new hire from this object. This is because it is the easiest object for the new hire to update in the system. Application data includes candidate information; however, it is a snapshot of when the candidate applied. Many things can happen between the time a candidate applies and when they are ready to be hired! Last names can change, degrees can be earned, residences can change, etc., so it is best to get this information from where the candidate can easily update and make sure the recruiters know to inform the candidate to keep this information up to date so that downstream systems get accurate data.

JobRequisition Object

The JobRequisition object is created when a user clicks the "Create Requisition" button within recruiting, a requisition is created from Employee Central position management, or an integration creates it from a third-party system. It contains information about a position such as the title, company code, and department. It also contains other information like the number of openings and the final time to fill once the requisition is closed. Additionally, information about the posting is associated with this object as well as associations with the recruiting team such as the recruiter and hiring manager. Typically, we see information about the position drawn from this object into HRIS systems such as the position ID that corresponds to the position ID in the HRIS system. Cost center information can also be pulled into payroll systems to understand where to charge payments to the employee. In cases where the pay is fixed based on the position, pay information can also be drawn from this object.

JobApplication Object

Anytime a candidate applies to a job posting or a recruiter adds a candidate to a requisition, a JobApplication Object is created. We tend to think of a JobApplication as the union of a candidate and a requisition (this can also help you understand why the system always requires users to create a candidate profile prior to applying to a posting). As mentioned in the Candidate object section, the JobApplication object takes a snapshot of the candidate profile at the time of application and stores it on the JobApplication. So it is important to take this in consideration when transferring data upon hire (e.g., Candidate data on the JobApplication may be out of date by the time you are ready to hire). JobApplications also make associations with other objects like assessments, background checks, interviews, etc. Additionally, in our experience, we tend to add custom fields to this object (defined in the template XML) that include important information about the candidate that is relevant (e.g., if the candidate is a rehire, whether they passed a rehire check, legacy system IDs, etc.). Items that we don't want to capture on a candidate profile until someone has actually applied tend to reside within the JobApplication object as well. For example, racial information, tax ID, and birthdate are sensitive data we tend to avoid collecting until needed and therefore often get collected on the Job Application toward the end of the recruiting process.

JobOffer Object

The JobOffer object is the object that stores data behind the offer approval screen. When you create an offer approval on an application, this object gets created. It is typically used to capture financial information for HRIS and payroll systems in the event that this information is not stored on the application or requisition. Since the JobOffer object is a child of the JobApplication object and the JobApplication is a child of the JobRequisition object, it provides the most specific information available (e.g., it is specific to a particular candidate rather than the entire requisition and is specific to the latest offer on the table). However, it might not be used in all scenarios (e.g., in the event of an hourly job that starts at a fixed hourly rate for all hired employees). In the most common scenario, a base salary and possibly any bonus information are sent to an HRIS/payroll system from this object. Sometimes a job start date, orientation date, etc., may also be included here.

Note   *In addition to not being required for all jobs, it is also important to keep in mind that multiple JobOffer objects may be associated with one JobApplication object (e.g., when a candidate rejects the first offer and a new offer is made). Thus, we recommend logic to be included in any interface that checks if the JobOffer object exists and, if one exists, then to pull information only from the latest one.*

Case Study: Sending Hire Data to a Payroll System Using a Custom CSV Format

Now that we've walked through the different types of data that are collected throughout the recruiting process and how they are stored, we are ready to start transferring relevant data to other systems now that you are ready to hire the candidate. We started the book looking at how requisitions could be created from a payroll system. To continue this scenario and since hiring into a payroll system involves significant data and is the most practical scenario, it makes sense to explore it further. Typically, payroll systems will want to import data in a predefined CSV format that cannot be changed. Being able to create a CSV file with the needed data without coding can help avoid middleware and developer costs. Let's see how it is done!

Note   *If you have already read through prior chapters, you are probably already familiar with using the Integration Center to gather the needed objects and create a CSV file. Here, we got into a little more detail about how to gather data from multiple objects and also change the format to the CSV file output to fit more rigid formatting needs.*

To begin our journey, let's pretend the HRIS system administrator has given us a sample CSV file as shown in Figure 13-2. We must output data in the same format as shown in Figure 13-2 in order for it to load into payroll. The first thing to notice is that this is a character delimited file instead of comma – and the character used to delimit data is a pipe character "|". When we create the file, we will need to make sure to configure it using this character instead of a comma, in addition to making sure the header titles and order of all of the fields we gather from throughout the system are in the same format.

Figure 13-2. *Sample Payroll Input File*

Note   *You would likely also have more fields on a payroll file such as type of pay (hourly/salary), work categorization, etc. We have trimmed this file down to show you how to get to different types of fields from different objects without being completely comprehensive.*

We will also need a trigger to know when to send the file and which applicants should be contained in the file. The simplest way to do this is to create an applicant status such as "Ready to Hire." Then, we simply use a batch job to periodically grab every applicant who is sitting in that status.

Note   *As a second step, we can have another job run to put the same people in a "Sent to Payroll" status or even in a "Hired" status if we get back a confirmation – the design of the second step would largely depend on the capabilities of the payroll system. For the sake of our example, we will skip this second step since we can't speculate too much on the specific capabilities; just know that these are viable options. You will only want to either manually or automatically put a person in the final "Hired" applicant status once you can confirm the hire is completed since it affects standard reporting in SAP SuccessFactors (such as time to hire and EEO reporting).*

Follow the given steps to create a CSV file following the format shown in the sample file in Figure 13-2:

1. Type and select "Integration Center" in the search box.

2. Click "My Integrations."

3. Click "+ Create" and choose "Scheduled Simple File Output Integration."

4. Since we are pulling from multiple objects, we will have to choose the most appropriate object that will give us access to the other objects where data is stored. Since the JobApplication object is connected to all other objects in question, we will start with this object. Type "JobApplication" in the "Search for Entities by Entity Name" field and click the top result on the left.

5. Next, we will select the fields we can from the JobApplication object and the navigation that will let us access the other needed fields from related objects. On the right, under "Fields," choose "First Name," "Last Name," "Address," "Address Line 2," "City," "Zip," "Gender," and "SSN." Under "Navigations," choose "Job Offer" and "Job Requisition" and "State." Click the "Select" button.

6. Now, we will make sure to use a pipe delimited format. On the Options screen, give the integration a name. Be sure "Output File Type" is set to "Simple Delimited," "File Delimiter" is set to "Pipe (|)," and "Header Type" is set to "Simple Header." An example is shown in Figure 13-3.

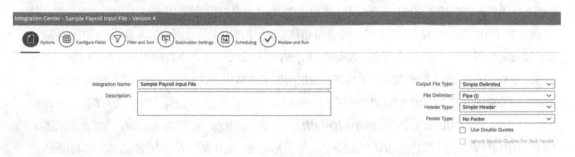

Figure 13-3. *Example Options Screen When Creating Sample Payroll Input File*

7. Next, we will want to add our main filter. Click "Filter and Sort" at the top. Click "Advanced Filters," and then in the "Field" field, click "Browse." In the new pop-up, click "Entity Tree View." This will allow us to navigate to related entities. Scroll down to "jobAppStatus" and click this object to expand it. Click "Application Status Id." An example is shown in Figure 13-4. Click "Change Association to Application Status Id" when you are ready to set the filter.

Figure 13-4. *Selecting "Application Status Id" for the Filter*

8. Next, enter the status ID for "Ready to Hire" in the "Value" column
    so that we only see applications in this status. An example is
    shown in Figure 13-5.

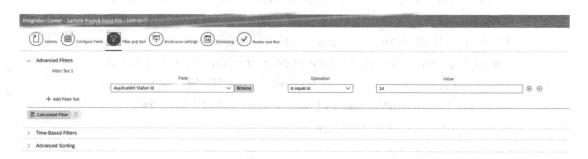

Figure 13-5. *Setting the Filter So That only "Ready to Hire" Applicants Are Shown in the Output*

Note   *You will need to look up the status ID in the Applicant Status Set. To do this, make sure you save your interface and then type and select "Edit Applicant Status Configuration." Click the status set associated with the requisition template for which you are exporting applicants. You will see the status IDs for each status within the status set on the left. An example is shown in Figure 13-6.*

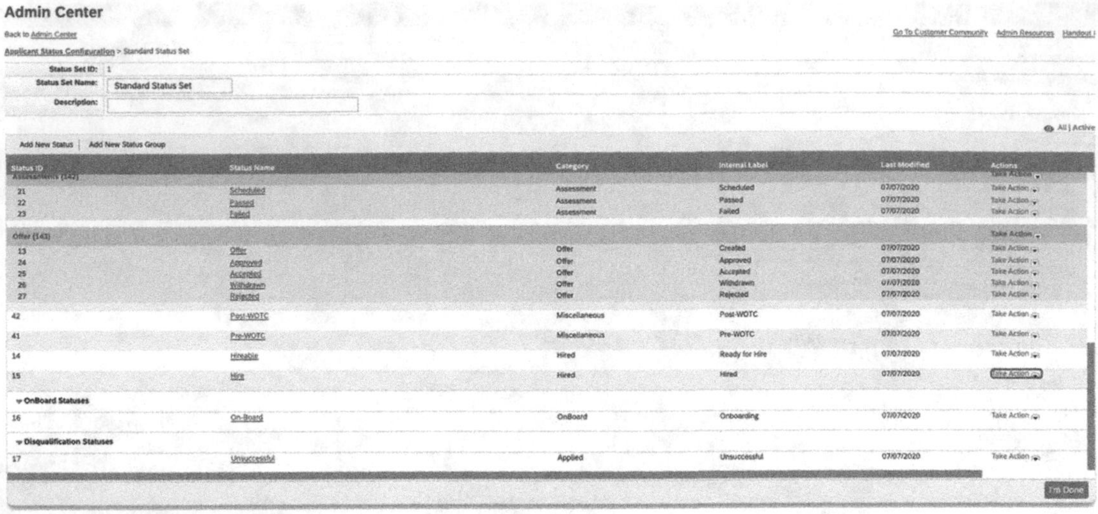

Figure 13-6. *Applicant Status Set Configuration Showing Status ID*

9. Next, we had one field on the jobApplication that is a picklist
    that could be translated to many languages. Since the payroll
    system file wants the spelled out name of the state in English, we
    will have to grab this translated label using a special sub-filer.
    Click "+ Add Field." Under the Entity Tree View, click "state," then
    picklistLabels, and then "label." Beneath, click the FieldFilters.
    Click "picklistOption" and choose <first> (since there will only be
    one option ever chosen, it is safe to default to the first). Then click
    "PicklistLabel" and choose "locale" under the "Field" column and
    type "en_US" in the "Value" column. Click the "Add Association"
    button. An example is shown in Figure 13-7.

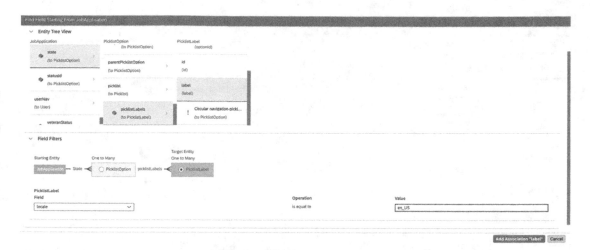

Figure 13-7. *Adding the Translated Picklist Option for the "State" Field*

Note   *More commonly, a payroll system will want a code. When your configuration specialist or system admin set up the picklist in picklist management, they could add a code for each picklist option instead as specified by the payroll provider.*

10. Now, we will add the fields we need from the other objects that have fields we need to include. Let's start with the cost center from the requisition. Click "Configure Fields" at the top of the screen."

11. Click "+ Add Field" in the upper-right-hand corner of the screen.

12. In the pop-up that appears, click the "Entity Tree View" and then choose "jobRequisition" on the left and then "CostCenter Code." Click the "Add Association" button. An example is shown in Figure 13-8.

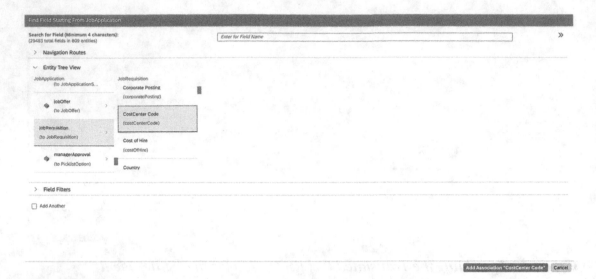

Figure 13-8. *Example Adding CostCenter Code from jobRequisition*

13. In an international implementation, you may also have your
data stored in different locales on the requisition. To add the job
title in the "en_US" (for English United States), click the "+ Add
Field" button. In the pop-up, under "Entity Tree View," choose
"jobRequisition," then "jobReqLocale," and the "Job Title." Click
"Field Filters" and then click "JobRequisitionLocale." Choose
"Locale" under the "Field" column and then type "en_US" under
the "Value" column. Click the "Add Association" button. An
example is shown in Figure 13-9.

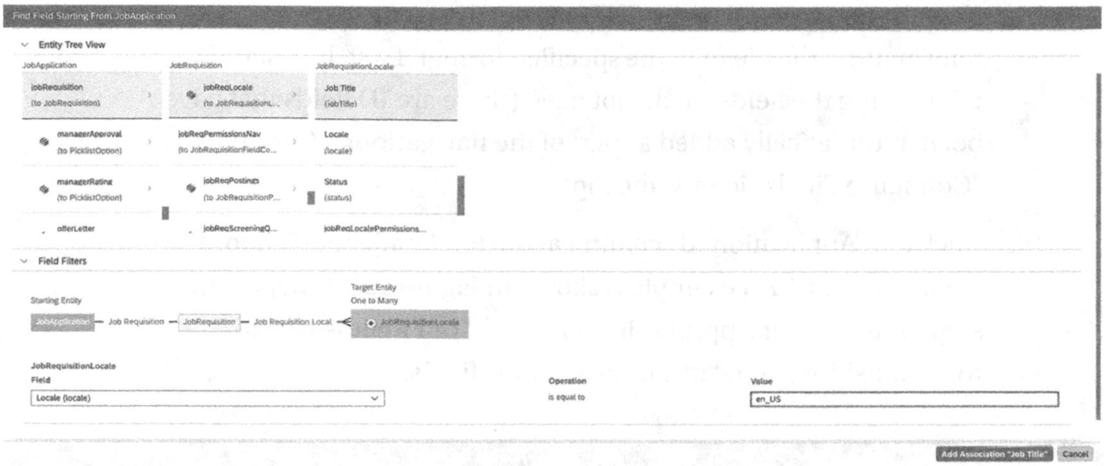

Figure 13-9. *Adding the Translated "Job Title" Field from the "jobRequisition" Object*

14. Next, we will add the salary field. Since we can have more than one jobOffer associated with a single jobApplication, we will need to also specify which jobOffer object to choose. Click "+ Add Field" again and then choose "jobOffer" on the left. Then choose "Salary Base" on the right. Click the "Field Filters" section on the bottom and then choose "<Last>" in the "Field" column. This will choose the last available JobOffer object created. Click "Add Association." An example is shown in Figure 13-10.

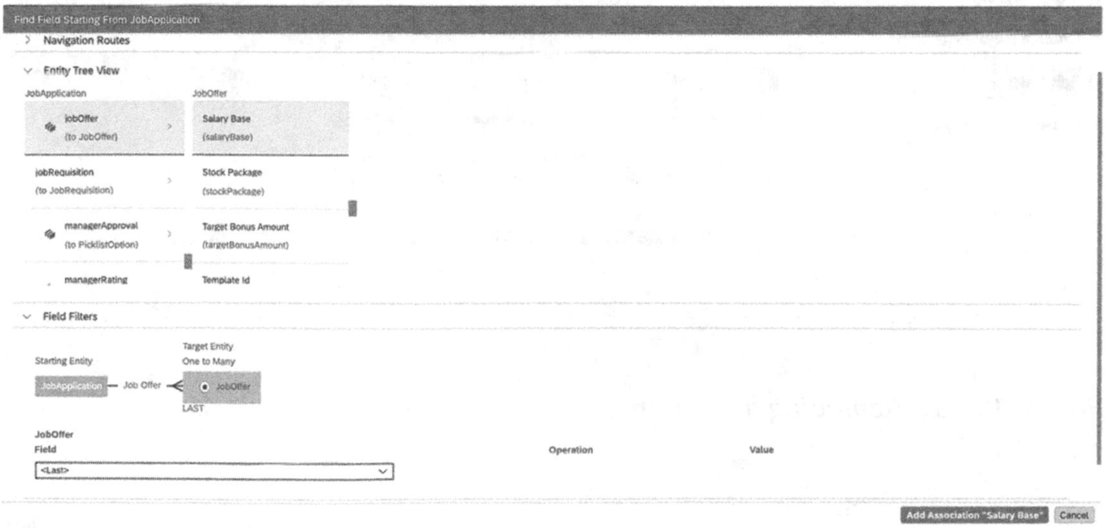

Figure 13-10. *Adding Fields from Job Offer While Applying a Filter*

15. Now that we have all of the fields we need, we can start reformatting the file to fit the specified format. First, let's start by eliminating the fields we do not need (there are ID fields that have been automatically added as part of the navigations). Click the "Configure Fields" icon at the top.

16. Click the "Application Id" column at the top header and choose "Remove Field." An example is shown in Figure 13-11. Repeat this step for the "Offer Approval Id-Job Offer," "Job Requisition Id-JobRequisition," and "id-PicklistOption" fields.

Figure 13-11. *Removing a Column*

17. Next, drag and drop each column header so that they are in the desired order of the sample file. An example is shown in Figure 13-12.

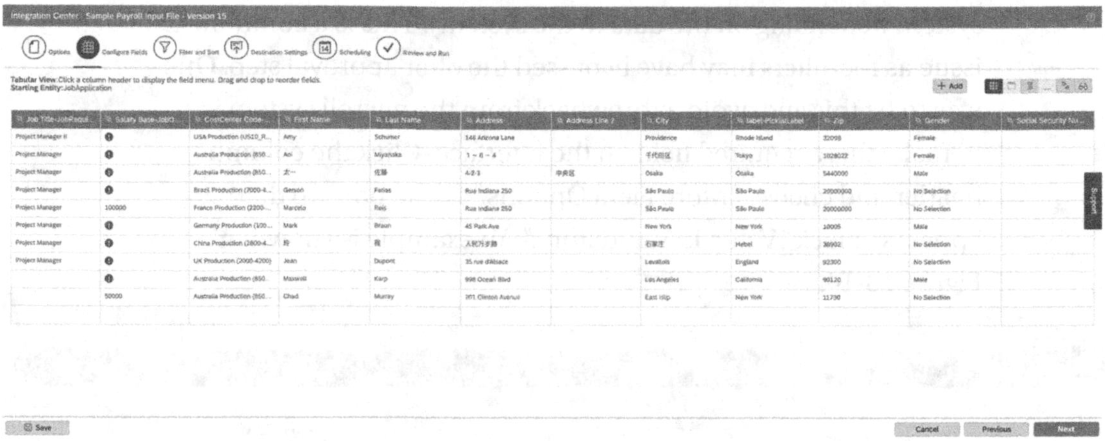

Figure 13-12. *Final Column Order Example*

18. Now let's change the headers to match what is in the sample file. Click each column, and in the menu, set the "Label" field to the corresponding label from the sample file and hit enter. An example is shown in Figure 13-13.

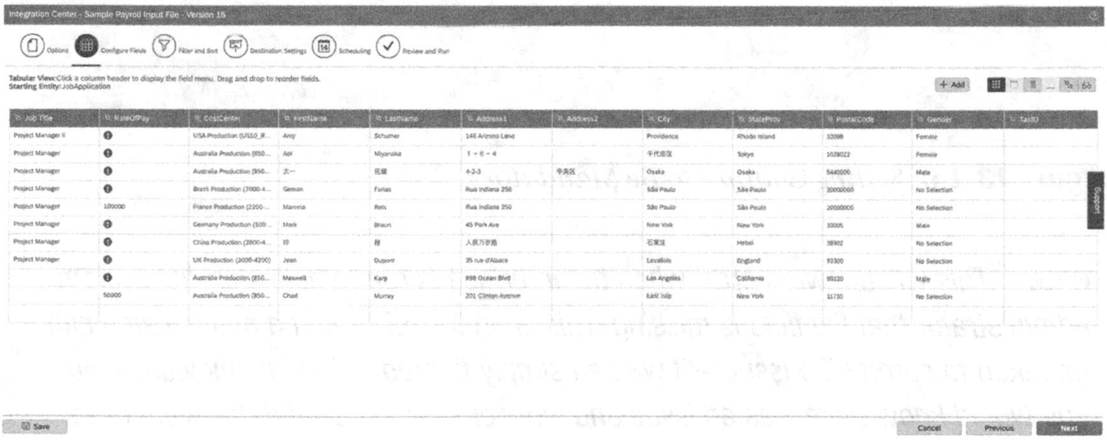

Figure 13-13. *Relabeling Column Headers Example*

19. We are almost there! Now let's look at the data formatting. The first observation when we look at this preview screen with the sample data in our particular example system is that the RateOfPay column does not have values for every entry (this will vary in your system depending on the data in the system). This is a common issue as recruiters may have bypassed the offer approval step. One way to fix this and avoid a throwback from the payroll system is to make this a required field on the interface. Click the column header and choose "More Field Options." In the pop-up that appears, check "Value Is Mandatory." An example is shown in Figure 13-14.

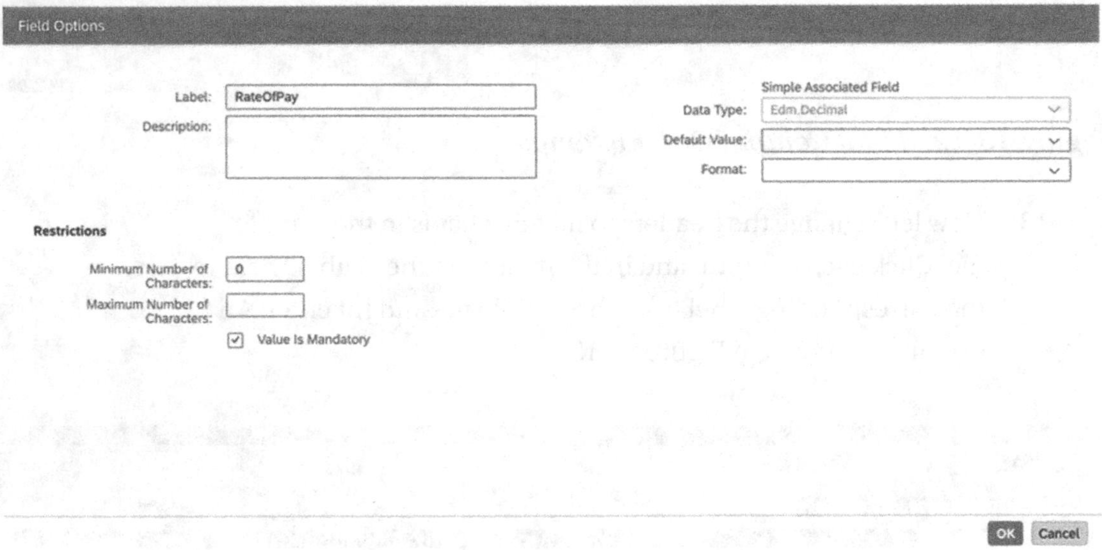

Figure 13-14. *Setting Column to Be Mandatory*

Note   *This will throw an error when the interface is run to help alert the system administrator that the field is missing. This is preferred so that a manual step can be taken to correct the issue – if we had simply filtered out the blank values, no one would know there was an issue and wonder why the candidates were never transferred to payroll!*

20.   Continuing to examine the example data on screen, the only
      item left is the cost center field and gender (you will notice
      TaxID is missing as well in Figure 13-13, but this is hidden in our
      screenshot for security purposes – we will assume the data is
      present). Looking at the sample data shown in the Integration
      Center screenshot in Figure 13-13, it seems perhaps the cost center
      codes in SuccessFactors have both a label and a code, whereas
      there is a different code in the payroll system. Similarly, "Gender"
      shows up as "Male," "Female," or "No Selection" instead of "M,"
      "F," or blank. One way to remedy this is to create a lookup table.
      Click the name of the column on top and then click "Value Lookup
      Table." In the pop-up, click "+ New." Give the table a name, and
      then for each value you want to convert, fill in the original value in
      the "Input Code" column and the converted value in the "Output
      Code" column. An example is shown in Figure 13-15. Click "+" to
      add more rows to the table and fill in the columns for all values
      you want to convert. Click "Save" when you are done building your
      table. Then click "OK." Repeat for the second column so that both
      "CostCenter" and "Gender" fields are remedied.

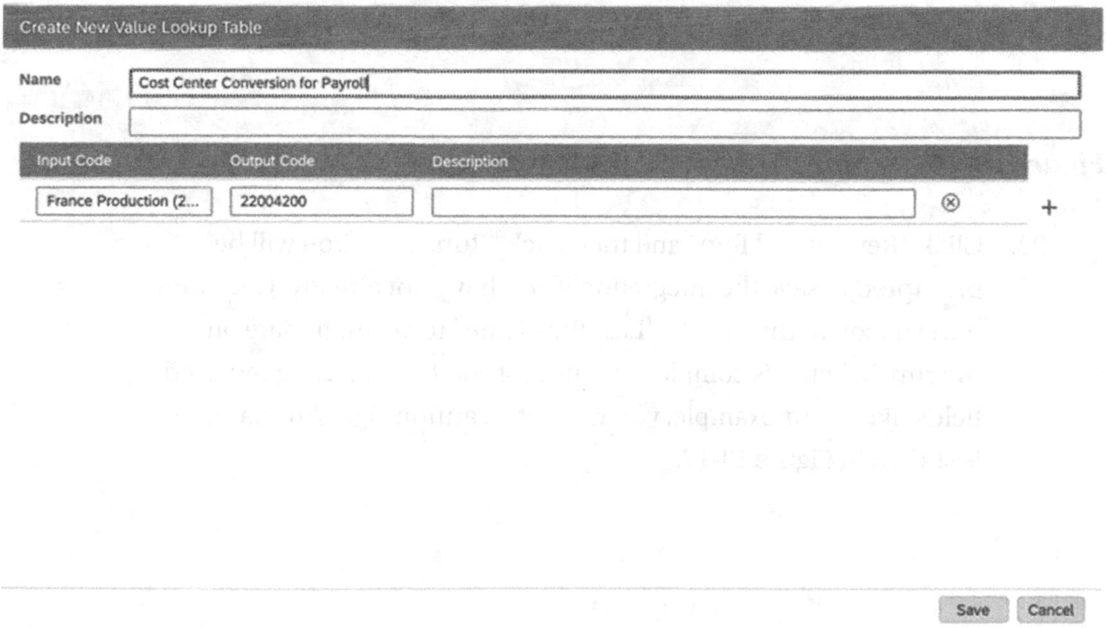

Create New Value Lookup Table		
Name	Cost Center Conversion for Payroll	
Description		
Input Code	Output Code	Description
France Production (2...	22004200	

Save Cancel

Figure 13-15. *Creating a Value Lookup Table*

21. Now that we seem to have an interface that complies with the sample format given, let's give it a test run. Click "Destination Settings" and fill in your SFTP destination settings on the left. On the right, let's pretend the payroll system is expecting the payroll file to have the name "PAYEMPINPUTyyyyMMdd. csv" where yyyyMMdd is the date stamp at the end (e.g., PAYEMPINPUT20210101.csv for January 1, 2021). An example is shown in Figure 13-16.

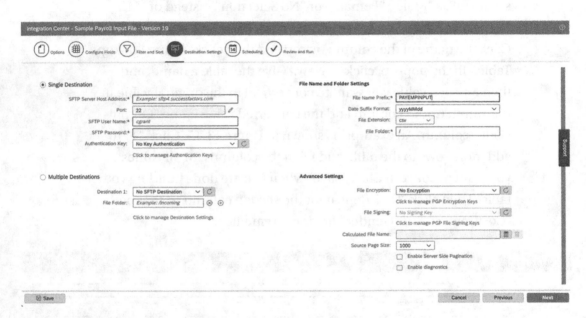

Figure 13-16. *Setting Up the SFTP Destination and File Name*

22. Click "Review and Run" and then click "Run Now." You will be prompted to save the integration if you have not already. Click the refresh icon to the right of "Last Run Time" to get an update on the run. When it is complete, assuming you have missing required fields like in our example, you will get a caution sign. An example is shown in Figure 13-17.

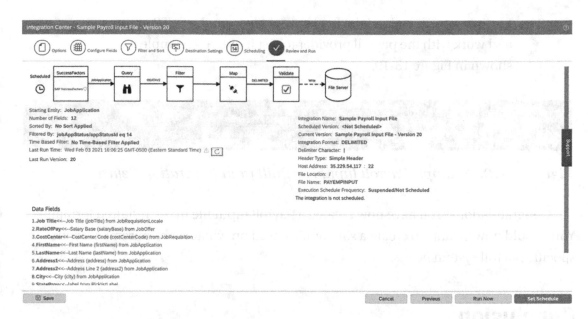

Figure 13-17. *Example Run of Interface with Warnings*

23.   Click the timestamp of the last run to see the details of the run.
In Figure 13-18, you can see the missing required values flagged
a warning. These rows were excluded from the output (which is
a good thing!), and the application numbers are provided so the
system admin can contact the recruiters and get the applications
corrected.

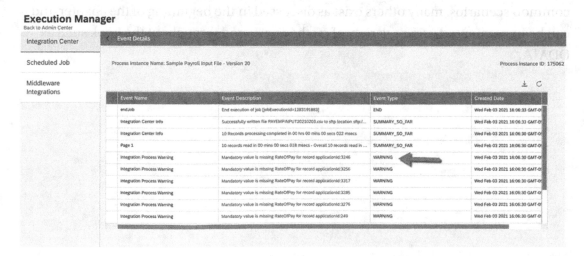

Figure 13-18. *Example Warnings Thrown for Missing Fields*

24. You can now download the final output file from the SFTP server and work with the payroll provider to test it out. An example is shown in Figure 13-19.

Figure 13-19. Example Payroll Input File Built from Integration Center

Congratulations! You have now created a payroll input file in the required format! You should now be able to create a similar file based on whatever requirements your specific payroll system needs.

Conclusion

In this chapter, we have focused on the final step in recruiting a candidate: hiring. We started by discussing some practical scenarios where recruiting data can be shared with various systems to help ease the hiring process. We then jumped into a review of the relevant data that is typically passed to other systems during the onboarding and hiring process. We wrapped up with a specific case study that walked through the specific steps we can take to set up an employee export in a specific format to be input into a payroll system. It is important to note that while this is probably one of the most practical and common scenarios, many others exist as discussed in the beginning of the chapter and can be implemented using a variety of methods covered throughout the book such as ODATA.

CHAPTER 14

Conclusion

We hope you have enjoyed your journey with us across each step of the recruiting process. Along the way, we've shown you various integration and automation opportunities as well as practical implementation scenarios that provide real business value. Let's take a moment to review what you've learned.

Review

In Chapter 1, we began by introducing ODATA. We learned conceptually what the ODATA protocol is. We also walked through how to make an example ODATA call. This prepared us for upcoming chapters which use this protocol and also set the stage for you to discover how powerful APIs can be in connecting systems to SAP SuccessFactors Recruiting and beyond.

As we continued into Chapter 2, we started our chapter-by-chapter progression through each stage of the recruiting process. Chapter 2 started us off with requisition update and creation. We started off providing some practical business scenarios where the need to update and create requisitions would occur. We began with a simple scenario of updating a requisition from a basic payroll system that acts as a company's HRIS. We then progressed to a slightly more complex scenario creating a requisition from a full HRIS system. We reviewed the relevant ODATA API objects before progressing into a walk-through of how to construct the files and/or ODATA calls necessary to make these scenarios work.

In Chapter 2, we alluded to the fact that sometimes middleware is needed to make these scenarios work. Thus, in Chapter 3, we began our journey into exploring how middleware can be used in even more complex scenarios requiring this technology. We reviewed what middleware is, when it is appropriate to use, and how to use it. We then stepped through a practical scenario of creating questions on a requisition automatically using this technology.

© Anand 'Andy' Athanur, Mark Ingram and Michael A. Wellens 2022
A. A. Athanur et al., *Innovative SAP SuccessFactors Recruiting*, https://doi.org/10.1007/978-1-4842-7425-5_14

Chapter 4 took us into the next step in the recruiting process which is posting requisitions out into the job market. In this chapter, we saw how the tools available in SAP SuccessFactors Recruiting allow us to repost to cloud content management systems like WordPress and walked you through a practical example.

As we continued in Chapter 5, we explored how to increase candidate engagement. We shared the strategy of creating a chatbot to engage candidates so that candidate contact information and interests were loaded into the system. We stepped through how to enter and retrieve information from the SAP SuccessFactors system via APIs to make a chatbot integration possible. In conjunction with standard SAP SuccessFactors candidate engagement functionality, we showed how this chatbot enhancement allows companies to engage potential candidates that may have otherwise left a company career site without any further engagement.

In Chapter 6, we looked at progressing candidates through the selection process. We examined Robotic Process Automation (RPA) as a viable technology to speed the progression and/or disposition of candidates through the system automatically in mass rather than manually clicking and selecting each candidate individually. This can be particularly useful for end users who do not have Integration Center access. We then continued to examine candidate progression in Chapter 7, this time using the Integration Center to speed the process. In both chapters, we took a look at practical examples and stepped through how to use the technology to automate candidate progression. After reading these chapters and seeing the practical examples provided, you should have some ideas of how to augment candidate progression in other ways.

Chapter 8 then took a deep dive into the Assessment Integration Framework. We learned how the framework provides a standardized way for vendors to integrate their candidate assessment systems in SAP SuccessFactors. We learned how assessment systems can help recruiters quickly sort through the most qualified candidates. Automating and integrating the connection to these systems can enhance this even further by making the assessment request and results seamless between systems. We continued examining this business process further in Chapter 9, this time looking at ODATA as a way to integrate with assessment vendors separately or in conjunction with the assessment framework.

Next, in Chapter 10, we examined the Background Check Integration Framework. We discovered how the framework is used as a standardized way for background check vendors to integrate into SAP SuccessFactors Recruiting similar to the Assessment Integration Framework. This can allow recruiters to automatically request and receive background check results to and from vendors without having to log into those systems separately.

In Chapter 11, we progressed to the offer stage of the recruiting process. You learned how offer approval forms could be automated using business rules. We took a practical example of providing an automated suggested offer amount and saw how this could be used in conjunction with other automations to create a comprehensive recruiting information system solution. By the end of this chapter, you should have a solid understanding of how to use a business rule to help automate the offer approval process. In addition, you could recognize how this business process could be automated using other methods and how business rules could be used in other business processes beyond offer approvals as well.

As we progressed into Chapter 12, we took a look at the onboarding process. Among the many available technologies to enhance this process (such as the SAP SuccessFactors Onboarding module itself), we took a look at Intelligent Services. We discovered how this platform-level feature of SAP SuccessFactors can be used to create service requests that can ease an employee's transition by having requested equipment ready for them on the day of hire.

Making our way into Chapter 13, we arrive at the final recruiting process: hiring. In this chapter, we examined how to automate hireing into a third-party payroll or HRIS system. We covered the data typically needed to conduct a hire and how to collect that data from SAP SuccessFactors. We also stepped into some of the more advanced features of the Integration Center to manipulate how data is sent to other systems.

Figure 14-1 shows us a depiction of each business process we have covered throughout these chapters along with the technology solutions used to enhance each. While we feel we've given a wide variety of technology solutions, this list is certainly not comprehensive. Indeed, new technologies are being developed every day!

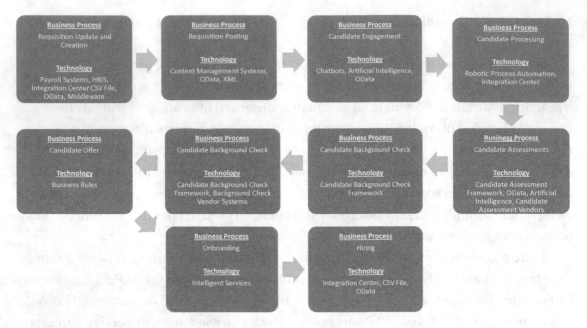

Figure 14-1. *Recruiting Businesses Processes and Associated Technology Solutions Covered*

Realizing Business Value

As we have covered each step in the recruiting process, we have strived to show practical case studies that bring real business value. While many of the benefits are obvious (e.g., less time needed to sort through qualified candidates or create background checks), the decision whether to go ahead with a particular automation or integration often depends on monetary business decisions. In these cases, it becomes necessary to build a business case before implementing such a solution. In addition, it is also important to measure the success of such an initiative after implementation to make sure the business value is being realized and adjust course as needed.

In an extreme case, if you have so little volume of candidates that you only need one recruiter to handle the entire recruiting process and this recruiter has extra time on their hands even after manually creating background checks in another separate background check system, then the expense of creating a background check system integration probably does not make sense. Conversely, if your organization is having trouble staffing multiple recruiters and they are working overtime to complete basic recruiting tasks, multiple automations make sense just to keep the company operational. Most organizations probably lie somewhere in between these two extremes. Thus, most companies probably need to create some measure within their business case to understand if the cost of creating the automation of integration is less than the cost of not.

When considering what measure to use for your business case, consider that automations and integrations are all about process efficiency. You are essentially reducing the amount of time it takes to perform a task. Thus, a typical way of putting a dollar figure around such a business case is to calculate a blended hourly rate for your recruiters. For example, if you have 10 recruiters who typically cost $75 per hour and they each spend 8 hours a week creating background check requests, then you are burning about $6,000 a week just in creating background check requests. If an integration reduces that to 1 hour a week, then you would save $5,250 a week. If the integration only costs $30,000 to implement, then you would make up the cost in less than 6 weeks!

There are also less-tangible benefits associated with many of the automations and integrations we included in this book, for example, candidate experience and recruiter experience. Indeed, there is a lot of focus in the industry now on user experience, with SAP dubbing SuccessFactors as "Human Experience Management" or "HXM." If candidates are impressed with the experience they receive while being recruited, they may be more likely to join and stay with the company. Similarly, if recruiters are frustrated with their processes or feel the company is technologically backward, they may be likely to leave the company. While there are KPIs that can help gauge pain points in the candidate experience and recruiter experience (e.g., satisfaction surveys, recruiter retention, and candidate drop rates), it can be sometimes harder to find the correlations between these trends and a specific automation and be confident of the exact causation of that correlation.

Taking all of these considerations into play, you should be well prepared to build a business case for implementing the types of automations and integrations we've covered in this book. In our experience, the benefits of these integrations and automations are quickly realized in real-life organizations.

Limitless Possibilities

In each chapter, we have covered sample technologies at each step in the recruiting process. However, it is important to understand that these are not the limits of possible integrations. For example, business rules are not limited to automating offer approvals. They could also be used to help automate requisition updates or application updates. ODATA calls can be used in just about any business process within and outside of the recruiting module in SAP SuccessFactors. Indeed, there is a near limitless combination of technologies and process steps that can be used to enhance the candidate and recruiter experiences as well as enhance the overall recruiting process. The purpose of this book has been to introduce you to some of those possibilities and engage your creativity. Now, go forth and create!

Final Conclusions

We hope you have enjoyed our technical journey throughout the recruiting process! Overall, you should now have a solid understanding of where opportunities to enhance and automate SAP SuccessFactors Recruiting exist at every step of the recruiting process. Furthermore, through the examples provided, you should now understand how to implement a variety of available technologies at each step in the process to make these opportunities a reality. In addition, you should be able to see how these automations and enhancements provide real business value through cost savings and enhanced recruiter and candidate experiences. Hopefully, the experiences and best practices we've collected and shared from our own careers have been of great value to you. We wish you well in your own recruiting systems adventures!

Index

A

B

© Anand 'Andy' Athanur, Mark Ingram and Michael A. Wellens 2022
A. A. Athanur et al., *Innovative SAP SuccessFactors Recruiting*, https://doi.org/10.1007/978-1-4842-7425-5

Printed in the United States
by Baker & Taylor Publisher Services

Printed in the United States
by Baker & Taylor Publisher Services